TEX

S. E. HINTON

Published by
Bantam Doubleday Dell Books for Young Readers
a division of
Bantam Doubleday Dell Publishing Group, Inc.
1540 Broadway
New York, New York 10036

If you purchased this book without a cover you should be aware that this book is stolen property. It was reported as "unsold and destroyed" to the publisher and neither the author nor the publisher has received any payment for this "stripped book."

Copyright © 1979 by S. E. Hinton

All rights reserved. No part of this book may be reproduced or transmitted in any form or by any means, electronic or mechanical, including photocopying, recording, or by any information storage and retrieval system, without the written permission of the Publisher, except where permitted by law. For information address: Delacorte Press, New York, New York 10036.

The trademark Laurel-Leaf Library® is registered in the U.S. Patent and Trademark Office.

The trademark Dell® is registered in the U.S. Patent and Trademark Office.

ISBN: 0-440-97850-5

RL: 4.8

Reprinted by arrangement with Delacorte Press

Printed in the United States of America

Three Previous Editions

November 1989

40 39 38

*for my grandparents
and David, one more time*

TEX

ONE

"There ain't no bear in that bush," I said. Negrito's ears were pricked so far forward they almost touched, and he was picking up his feet like he was walking on eggshells.

"You've never even seen a bear, you dumb horse," I told him, keeping a strong leg on him. "You don't even know what one looks like."

Negrito blew through his nose, rolling one eye toward the bush.

I laughed. Negrito had a good imagination.

A sudden gust of wind rustled the bush and Negrito gave a snort and a huge sideways leap and tried to take off. I kept him back long enough to let him know it was my idea to gallop, then we went. The drumming of his hooves was better than music.

"There's a ditch coming up, man," I said. I

could tell by the way he was holding his head up that he saw it.

"Do you want to jump it or not?"

I felt him weave a little, so I leaned back and pushed, tightening my leg. "Sure you do."

We soared over the ditch without breaking stride, but once we landed, he bobbed his head and bucked a little.

"Boy, that was fun! You're a great jumper. A really great jumper," I said, slapping his neck with the reins. "Next year we make the Olympics."

Then, worried that I sounded too sarcastic, I added, "For a cow horse you are a really good jumper."

We loped on. This was the first day I could really feel fall coming on, not so much because it was chilly, but there was a slant to the sunlight and a smell in the air that meant fall.

Pop ought to be coming home pretty soon. Summer, shoot, there were lots of rodeos going on, lots of places he could be all summer, but fall would be a really good time for him to come home.

Negrito jumped sideways and started bucking again.

"Geez, it's a rabbit. For Pete's sake, don't you know a rabbit when you see one?"

Negrito shook his head. I got him collected till was bunched up like a coiled spring and his canter felt like a rocking chair.

He was just playing around. He was a pretty brave horse, actually. Fall always made him feel good. Besides that I hadn't ridden him for a while. My best friend, Johnny Collins, got a motorcycle for his birthday a month ago, and I'd been spending a lot of time dirt-biking with him.

I slowed Negrito down to a walk to cool him off. I had to get back and change clothes before I went to school, and I couldn't leave him hot. He kept breaking into a jog trot. Fall mornings he could go forever.

"I've been wasting a lot of time with that cycle," I said, while I was unsaddling him. "But it was new and everything and Johnny kept pestering me to go with him, but we'll go out for rides more."

Negrito turned and nipped at me. Sometimes he meant it, but mostly he'd just catch my sleeve or my jacket. I slipped the bridle over his ears. He almost knocked me down trying to scratch his head on my arm. The bit always made his mouth itch.

"Seeya later." I swung over the fence. Negrito stood there, waiting.

"Okay." I pulled the last piece of carrot out of my pocket and gave it to him. Then I just walked off because I never could convince him that I didn't have any more carrots.

Across the paddock my brother Mason's horse, Red, stood swishing flies, looking bored out of his mind. Mason had never treated him like a person, so Red had never acted like one. Mason was a pretty good rider, though. Not as good as me. Even Pop admitted that. A couple of years ago, when me and Mason did junior rodeos a lot, I always won more.

I barely had time to change my clothes before Johnny drove up on his cycle. It was a lot better way to get to school on than the bus. Mason always left early for school, so I hardly ever could catch a ride with him.

I hopped on the cycle. Riding double was against the law, since neither one of us was sixteen, but you know, that was kind of a dumb law. I don't go around trying to break laws, but I couldn't get real worried about that one.

It was one of those long days at school. When it was over, I met Johnny.

"Straight home today," I told him.

"Aw, come on, Tex, I was going to go to the gravel pits today."

"No way, man. This is a perfect day for horse riding."

"Listen, man, I'll let you drive." Johnny tried bribing me.

"Nope."

"Well, hang on." Johnny did a wheelie leaving the school grounds, and the principal saw him do it. I knew we'd both get sent to the office tomorrow. I get sent to the office quite a bit. Even more than Johnny.

"You can't do a wheelie on a horse," Johnny shouted over the engine.

"That's true," I yelled back. He was right there. You can teach a horse to rear up, but that is the worst dumb trick you can teach a horse.

Lord knows enough of them come up with it on their own. One quick way to get killed is having a rearing horse come down backward on top of you.

"Hang on, I'm going to jump the ditch!" Johnny shouted. I tightened my grip on his belt. The cycle flew through the air, bounded, skidded, and slid to a sideways stop in front of our house. I swung off the back end.

"Thanks for the ride."

Johnny took off his helmet to wipe the sweat off his freckles. Everybody else in his family was either real dark like their mother or real blonde like their father, and there Johnny sat, flame-haired as a matchstick. He always said he was a throwback. A throwback is like when you breed a chestnut to a black and get an Albino colt like its great grandfather. I read a book once where that happened. All the Collins had dark blue eyes, though. It was their trademark.

"You sure you don't want to go to the pits with me?"

"Not today. All I want to hear on a day like this is hoofbeats. Negrito is going to love a good run."

"Now just who is your best friend—me or that horse?"

I looked at him for a minute, thinking about it. "Well, let me see . . . I've known you both about the same amount of time . . ."

Johnny belted me in the stomach with his

helmet. "Hey," he said suddenly, "what's Mason doing home?"

I looked at our beat-up pickup sitting in the driveway.

"Huh. Maybe he decided not to practice today. It's a little while till the basketball season's really going, anyway."

"But him and Bob were meeting every afternoon to shoot baskets."

"He's probably out jogging. Mace ain't making a move that doesn't have basketball behind it."

"Shoot," said Johnny. "If I could get the city paper to come out and take pictures of me, I might give up this machine and jog myself."

"Wouldn't do you any good. The coaches won't take midgets."

Johnny rapped me with the helmet again, but he wasn't mad. If he minded about being short I wouldn't tease him about it, but nothing much bothered Johnny. I'm the same way.

"You ought to see our room," I said. "Ol' Mason has the walls papered with that picture. Come on in and have a Kool-Aid."

Johnny shook his head. "I want to get some biking in before I have to go home. I'm trying to keep it down to just an hour late. Cole's started making noises about taking my cycle keys away."

All the Collins kids called their parents by their first names. They were the only kids I knew who did that.

Johnny put his helmet back on. "Seeya, Texas."

He popped another wheelie going across the yard.

"Way to go!" I hollered after him. I went on in the house to change jeans before I went riding. I kept my jeans sorted by clean, sort of dirty, and real dirty. It saved trips to the laundry. The ones I had on could go a couple of days as sort of dirty, so I needed some real dirty ones to ride in.

"Mace? You home?" I yelled from the bedroom. The whole house was quiet.

The next place I headed was the kitchen. We have a real little house, white wood frame, a front room with one bedroom on the side, a kitchen right behind the front room. The bathroom is through the bedroom. There's an attic bedroom upstairs, but it's so hot in the summer and cold in the winter that Pop never uses it when he's here, but puts a sleeping bag on the living-room sofa.

I was hunting through the icebox for something to eat, when I saw something just out of the corner of my eye. I almost jumped halfway across the room. Turned out it was just Mason, sitting quiet in the corner, behind the table.

"Boy, you sure spooked me. What you doin' there?"

"What does it look like?" Mason is a pretty sarcastic person. I don't pay any attention. That's just the way he is. He sounds meaner than he is.

"We got anything to eat?"

"There's some baloney left."

I found the baloney and a jar with some mustard left in it and pulled a chair around to face him. "What you doin' home so early?"

"I cut school today."

I stopped scraping the mustard jar, astounded. "You kiddin'! Mr. Super Study cuttin' school? I guess that makes it okay for me to skip a day."

"I guess it don't." Mason is seventeen, a couple a three years older than me. Most of the time he seems even older than that. He's got into the habit of bossing me around. I don't know where he gets that. Pop never bosses us around. I just let Mason rattle on and don't pay much attention. That's just the way he is. He's always hollering or preaching about something.

"You go fishing?" I asked. Mason loves to go fishing. That's about the only time he unwinds. He didn't have much time for it anymore.

"No. I didn't go fishing." He was looking real strange, sitting there behind that table. He was too quiet. Neither one of us is quiet people.

"I thought I told you to stay off of Johnny's cycle." He didn't have any expression in his voice.

"No you didn't. You said I'd get myself killed ridin' Johnny's cycle. You didn't say to keep off it."

"Do I have to write everything down in letters three feet high and shove it under your nose? I wouldn't ride in a *car* if Johnny Collins was driving."

Now see, Mace says stuff like that he don't

mean. He liked Johnny all right. Johnny's older brother Bob was Mason's best buddy. Even their sister Jamie thought of us as a couple of extra brothers, even though she already had four. He knew I'd been riding with Johnny ever since he got the bike. He just decided now would be a good time to fuss about it. I just went on spreading mustard on the baloney and eating it. We were out of bread.

"We get any mail?" I set the empty mustard jar down on the table.

"Who'd be writing to us?"

"Pop."

"He never writes."

"Sure he does. We got a card not too long ago."

"Four months ago."

"Yeah? It don't seem that long ago. He's only been gone five. I guess he'll be coming home pretty soon now that the rodeos are mostly over. Maybe I'll go with him next year."

"I wouldn't count on any invitations, if I were you," Mason said distinctly. It was true Pop never had asked us to go along, but shoot, maybe he would, next year.

I looked at Mason again. Something about the way he was sitting there made me think he'd been like that for hours, just sitting and staring.

"It sure is cold in here." I was beginning to wish I'd left my jacket on.

"A guy came out and turned off the heat this morning. I haven't paid the gas bill lately."

"No kiddin'? Why not?"

Mason just looked at me like I was the dumbest thing on earth.

"We run out of money?" I asked. "What happened to all that money we made last summer?"

Mason gave me a real mean look, almost like he hated me. It gave me a jab in the gut, even though I was pretty sure he didn't mean it.

"I blew it all in my fun-filled week at Acapulco. Where do you think it went? Food, gas, clothes. How far do you think a couple of hundred bucks stretches? Maybe if you took paying customers instead of mowing lawns for free . . ."

I ignored that. I only did a couple of lawns for free and everybody else paid me—Mason likes to make a big deal out of little things. I've just learned to live with it.

"Well, what we doing about it?" I asked him.

"We," he said, more sarcastic than ever, "have already taken care of it. The heat will be back on tomorrow. *We* don't have to worry about it anymore."

When Pop was gone Mason took care of all the bills and business stuff. I didn't know nothing about them. He did. I never thought it bothered him. He was the kind of person who liked to run things, anyway.

I decided to change the subject. Talking about money always got Mace irritated. He hates being poor.

"I heard you broke up with Laurie."

Mason was a very private person. He was a

fanatic about keeping his personal life personal. Unfortunately you can't be the school basketball ace and keep your personal life personal. I figured if the whole school knew something, there wasn't any reason why I had to play ignorant.

"She was getting too serious. Talking marriage."

"Yeah? Scared you off, huh? You'll end up married one of these days, though."

"Not in the near future. Anyway it's none of your business."

"Boy, somebody sure put a burr under your blanket! Anyway Lem Peters says bein' married ain't so bad once you get used to it."

I wouldn't mind Mason getting married, to tell the truth. At least that'd take his mind off college. That was all he thought about, college and how to get there. Unless Pop came home, I wasn't too crazy about him going off to college and leaving me here by myself.

"Lem Peters better like being married. He's going to be stuck with it for a while."

"Just because you have to do something don't mean you can't like it, too. You know he wanted to get married besides having to." I glanced at the clock. "I better go get the horses rounded up. It's goin' to be too dark to ride pretty soon. Anyway if they ain't fed on time they'll be tearing the fence down."

I was halfway to the kitchen door when Mace said, "They ain't there."

"Yeah?" I said. "Where are they?" I thought

maybe he had turned them loose in the next pasture, even though since Cole Collins was leasing it for his cows we weren't supposed to keep the horses in it. We'd been running low on hay lately, though, with most of the grass in the half-acre paddock gone.

"I sold them," Mason said. I just kept looking at him, waiting for the punch line. I knew he didn't sell Negrito.

"No, you didn't," I said finally. He was lying or kidding or crazy. I was getting a sick cold feeling.

"Yeah, I did. Got a good price for them, too."

I didn't believe him. He couldn't sell Negrito any more than he could sell me. But just to make sure I ran out the back door, jumped down the back steps, and raced out to the barn. It was just a little lean-to, really. It'd been part of a barn once, but the rest of it had kind of fallen apart.

If I was ever late feeding the horses, they'd start trotting up and down the fence, nickering. Nothing made Negrito madder than being fed late. He'd pace the fence, his head getting lower and lower till his nose would be practically dragging on the ground, then when he saw me he'd paw and stamp and say, "Where the hell have you been?" He had a real thing about his food.

Neither horse was in sight. I whistled. Nobody answered. I ran up the little hill that led to the Collins' big pasture. Even across thirty acres I can tell horses from cows. They weren't in the pasture.

I felt really strange, like I'd swallowed a block of ice and it was just sitting in my gut, sending cold waves all over me. My head felt spacy, almost like it was going to go floating off my body. I walked back to the house. I was breathing funny. I couldn't get enough air.

"Mace," I said. He still sat behind the table, like he hadn't moved a muscle since I left, and he didn't even blink now. I gave him one last chance. "Where's them horses?"

"I told you I sold them. I don't know why you'd think I'd start lying to you this late in life. Don't you ever close doors? No wonder I can't pay the gas bill."

"Who'd you sell them to?"

"I won't tell you. Nobody you'd know, anyway. They've got good homes. I made sure of that. They're gone. Shut up about it."

I was walking around and around in circles. I couldn't think. I couldn't breathe. I started shaking and sweating like a horse being reined in and spurred at the same time.

"Mason!" My voice shot up a note or two. "You better get those horses back! I mean it, man!"

Mason just sat there and didn't move. It was like seeing a stranger wearing a mask of my brother.

"I ain't gettin' them back." He spoke softly, his teeth clenched. "They're gone and they're gonna stay gone. We couldn't feed them through the winter. I wasn't going to watch them starve to death. So just shut up."

"You better get those horses back!" I shouted. I picked up one of the jelly jars we used for glasses and slammed it against the wall. Mason jumped a little when it shattered, like he hadn't expected it to break. I couldn't stop moving. I grabbed the mustard jar and hurled it at the window. It crashed through the glass and knocked the screen loose.

"Texas," Mason said, "you better quit it."

"I WANT THOSE HORSES BACK!" I grabbed another jar and smashed it into the sink, breaking a few dirty dishes.

Mason shoved the table out of the way and came charging across the room. My mind went into a white-hot blank and I went crazy.

All I wanted to do was kill him. And even though I was landing a few punches, I didn't seem to be hurting him, which made me madder. We rolled on the floor, through broken glass. I was out to get him any way I could, biting, kicking, screaming, and cussing. He got me pinned by one shoulder and slammed his fist into my face. It was like getting kicked in the head by a horse. I couldn't see for the yellow sparklers in my head, and he let me have another one. For the first time I realized he was as mad as I was and crazy enough to kill me. I blocked the next punch with my arm and turned my head quick as a snake and bit into his other arm. I set my jaw and wouldn't turn loose.

Mason was swearing and slamming me across the face, and when that didn't knock me loose, he rolled off and brought his fist down hard in the middle of my stomach. I gasped for air, doubling up. I couldn't move. I lay there, waiting for him to finish me off while I couldn't breathe.

Mason just sat there, holding his arm, panting. When I was sure he had quit the fight, I turned over and lay face down, my head resting on my arm. I was hurting real bad. All my energy was gone. I think if the house had been on fire I wouldn't have had the strength to leave. I was crying and too tired to keep from it.

"Tex?"

I didn't move. He took hold of my shoulder, easy, to let me know he wasn't fighting any more, but I jerked loose.

"Look, I didn't mean to . . . dammit, Tex, I just want to see if you're hurt."

"I ain't hurt." My voice sounded weird, I think it was because I couldn't breathe through my nose. I stayed still, waiting to get my strength back, trying to make some sense out of what happened. I couldn't figure out what had happened.

"Lookit, Tex, it wasn't you—I mean, I didn't aim to take it out on you like that."

I didn't know what he was talking about and I didn't care.

"I'm going to get my horse back," I said. "Pop wouldn't let you sell those horses if he were here."

"Pop isn't here!" Mason shouted. "Can't you get it through your thick skull that all this happened because Pop isn't here!"

I flinched a little. For a second I thought I felt his hand on the back of my hair, then he muttered, "Well, hell."

Pretty soon I heard the back door slam and the pickup engine start. He'd go drive up and down the highway for a while. He always did that when he was mad.

I couldn't seem to stop crying. I cried because I was hurting and because I wanted to kill Mace, and he was my only brother and I didn't really want to kill him. I cried because Mason had never beaten me up before. Mostly we got along pretty good. Finally I thought about Negrito being gone and Pop being gone and I bawled like a baby. I never cried much before and I wasn't used to it and I didn't know how to stop.

I sat up after a while, wiping the tears off my face with my sleeve. Blood came off with them. A back tooth had come loose so I went ahead and pulled it. It was one I was going to lose anyway.

The kitchen looked like a wreck, with the broken window and shattered glass all over, the table turned over and blood splattered around. It was a real mess. I didn't like looking at it. I got to my feet. I hurt all over. When I was helping Lem Peters break his Appaloosa colt, I got thrown probably ten times a day, and I never felt this bad. I was shaking. I stumbled to the john to look in the mirror.

I looked like Donald Duck. My nose and mouth were swollen, my lower lip cut. I didn't look much better after I washed the blood off. One side of my face was darkening to purple. My left eye was swollen shut. If it'd been Halloween I could of got by without a mask.

I wanted to change to a clean shirt, but I didn't have one, so I changed to one that at least didn't have any blood on it. It smelled like horse, and almost set me off crying again.

I walked around the front room for a few minutes, slowly. It hurt to be moving, but I couldn't sit still. I kept trying to think what to do next. Finally I put on my jacket and started down the road.

It was getting dark. Once in a while a car would whiz by, but nobody stopped. Mostly it was quiet, except for a whippoorwill that I finally left behind. I walked as fast as I could, but once in a while I'd get a pain wave that'd slow me up for a second.

I heard the cycle coming up behind me, but I was thinking about other things. When it stopped and a voice said "Tex?" I couldn't even remember who I knew that had a cycle.

"Where you goin'?"

I stopped and stared at Johnny and his sister Jamie like I'd never seen them before.

"Good Lord!" Jamie swung off the back end

of the cycle to take a better look at me. "What happened to you?"

It seemed like so much time had passed since I'd seen Johnny last, I was surprised he didn't look older.

"Tex?" Johnny hopped off the cycle to push it alongside me. "What's wrong?"

I kept on walking. "Mason sold the horses and beat me up, and I'm gonna find them."

"Tonight? You know where they are?"

"Nope. But I'm gonna keep looking till I find them."

A car drove by and slowed down curious-like, then drove on. Johnny and Jamie practically had to trot to keep up with me. I'm pretty long-legged.

"Tex, you're being crazy. You can't just walk around the countryside till you find the horses," Jamie said.

"Wanna bet?"

"Your front door is open and you left the lights on," Johnny said, like that was real important. I could tell they thought I'd flipped out.

"You're going home later, though?" Johnny asked. "Tonight?"

"Not till I find the horses."

"Look, Tex, you can't sleep out here on the road!"

"I ain't sleepy."

"Listen," Johnny said, "I'll take Jamie home and come back for you, okay?"

I turned to face them. "Can't you two understand English? I said I'm lookin' for Negrito! Now get lost!"

They were shocked. I'm not a bad-tempered person mostly. In fact mostly I'm real easygoing. Standing still was making me shiver again, so I started on.

"Tex . . ." Johnny gave it one more try.

"Johnny, wait . . ." Jamie said. They had stopped and I was too far down the road to hear what they were saying. Pretty soon the cycle started up again and headed back in the other direction.

I was glad they were gone. I liked them both, but I had things to think about.

Every time I passed a pasture with horses, I'd stop and look. I'd know Negrito in the dark. But I was pretty sure Mason wouldn't have sold them to anybody close by. He knew I'd just go get them back. Well, it didn't matter where they were. I'd still get them back.

Another car came by, slowed down, and stopped. I looked over, figuring to tell the driver I didn't need a ride, when I recognized the truck. I started running. The door slammed. I hadn't got far when Mason pulled me up by my jacket.

"Get in the truck," he said. My heart was pounding so hard I could barely hear him.

"No," I said.

I was shaking, not from being scared of him, just shaking for no particular reason. Mason must have thought he was causing it, because he let go of my jacket. But when I didn't move, he said, "I'll tie you up and throw you in that truck, Tex."

I can't stand being tied up. Even when I was a little kid, playing cowboys and horses, I couldn't stand being tied up. It made me sick. Mason knew that. He wouldn't do that to me. But I'd never thought he'd pound on me, either.

I walked past him to the truck. I wasn't about to sit up front with him. I climbed into the truck bed and leaned back against the cab. I'd thought I was all cried out, but tears kept running down my face. The wind dried them up real quick, though.

When we pulled into the driveway, I hopped over the side before the truck stopped. Johnny's cycle was parked in the yard. I figured he and Jamie would be inside, waiting to see if I got home. I had one last thing to say to Mason.

"You can't keep chasing me down all the time. I'll just go looking for them again tomorrow."

Mason came around to my side of the truck. Apparently he stopped at the Safeway while he'd been driving around, because he started hauling out sacks of groceries.

"You can leave for Timbuktu tomorrow for all I care," he said. "But you're not gonna find the horses and you're not gonna give me any more trouble tonight. Now shut up and take some of these sacks."

Now while Mason had always been a sarcastic person, he'd never been out and out mean before. It was hard to believe what I was hearing. I almost told him what he could do with the sacks. Then a bruise on my face gave a big throb, and I decided to keep quiet. I took a couple of sacks of food and went on in the house. I hadn't seen that much food all together at one place for a long time, and I wondered where Mason got the money. Then I remembered, and almost dropped the bags.

Johnny and Jamie were sitting on the floor in the front room, playing cards. Johnny jumped up to push open the door for me. His mouth fell open. I reckon I looked a lot worse in the light.

"Did Mason . . . ?" his voice trailed off as Mace came in behind me. I marched to the kitchen, slammed the bags down on the counter, and marched out, kicking an overturned chair out of my way. Thank God for the Collins. I wasn't looking forward to being alone in that house with Mason.

I sat down on the floor with them. "Deal me in."

"I think you ought to call the cops," Jamie said. Our house is so small Mason could probably hear her even in the kitchen. "Have Mason arrested for assault and battery."

I just shook my head. I stared at the hand I was dealt, kept a three, and threw away a pair of jacks. I wasn't exactly in prime poker form.

I could hear Mason cleaning up the broken glass and stuff in the kitchen. I couldn't tell if I felt worse about him or losing Negrito, and I swear, if the Collins hadn't been there, I would have started in bawling again.

Footsteps sounded on the front porch, and Bob Collins, Johnny and Jamie's older brother, stuck his head in the door.

"Hey are Johnny . . . Tex, what happened to you?"

He came on in. I didn't look up. Jamie said, "Why don't you ask the child abuser? He's in there." She jerked a thumb toward the kitchen.

Bob looked at me uneasily. "Mason didn't do that?"

"Robert!" Mason called from the kitchen, "come here a second."

Sometimes Mason called Bob Robert. He was the only person who did that. Bob didn't like to be called Robert. He took a lot off Mason, though.

"By the way," Bob called, "Cole is looking for you two. I have the feeling there's going to be some butts blistered."

"Aw, Bobby, why didn't you say so?" Johnny frowned. His father would still whomp him one if he thought Johnny needed it, and hardly a week went by that Jamie didn't get spanked or grounded or both. Bob, being seventeen, could get away with more.

"Let's get going," Johnny said, picking up his helmet.

"Oh, we're going to get it, anyway, so we might as well stay."

"If Cole finds us *here*, we're dead ducks."

Cole Collins didn't like his kids hanging around me and Mason. He thought we were bad influences. He might have found out Pop had been in prison once, too, but I wasn't sure. I never told anybody, not even Johnny. Mason said he'd skin me if I did. But Cole might have found out, anyway. There were a lot worse people than me and Mason, but you couldn't tell Cole that. He judged a lot by money, and we sure didn't have any.

"I want to make sure Mason's through beating up on people for the night," Jamie said. I looked over at her and realized I liked the way she cut her hair. It looked like somebody had put a bowl over her head and cut around it. She had almost-black hair. It looked good with blue eyes.

"What could you do about it if he isn't?" Johnny said.

"I don't like my friends getting pounded on."

"He's *my* friend," Johnny said.

"*My* friend, too," Jamie said.

I felt kind of dumb, sitting there being talked about like a dog or something. I was real tired, too. I felt like I had a headache all over.

"Maybe you better go home," I said. "I don't want you guys to get into trouble."

"Yeah, Jamie, if we get grounded this week we can't go to the Fair."

"Well, okay. Just wait a minute." Jamie went to the door of the kitchen. "If you ever hit Tex again I'll . . ."

I don't know what she was going to threaten

him with. Mace just cut her off short. He said something he doesn't usually say to girls, but with four brothers I doubt that it was anything Jamie hadn't heard before.

And Bob said patiently, "Jamie, go home."

"All right. Come on, Johnny. Let me drive, huh?"

"Cole said you weren't supposed to."

"Cole," she said sarcastically, "is not here. You don't exactly knock yourself out to obey his every rule, anyway."

"Girls . . ." Johnny began, and Jamie snapped, "You give me any of that garbage and I'll make you sorry."

Johnny gulped in mock fright but handed her the keys and the helmet. Jamie had the reputation of being a really mean person. "Bye, Tex. Seeya."

I just nodded.

"Come on," said Jamie, "I bet I can hit sixty between here and home."

They only lived about a half a mile away.

"You better not," called Bob. Big brothers are all alike.

I almost felt like going to bed, I was so tired and achey. But it wasn't time to go to bed, so I picked up a *Western Horseman* magazine off the floor and crawled into the big chair to look at it. That put me right next to the kitchen wall and I could hear everything Mason and Bob were saying. I hadn't planned it that way, but I didn't move, either.

". . . swear I could have killed him," Mason

said. "It was like I blew a fuse or something. And it wasn't his fault. Damn Pop—he's the one who ought to be punched out. God, I didn't mean to hit Tex like that."

"Well, he looks like he'll live. Forget about it, Mace. He will."

"I'm not so sure. I wouldn't."

"You're not Tex. Mason, we could have bought the horses. You know Cole is already looking for a way to get Johnny off that bike, and Jamie's going to start hounding him for one pretty soon."

"Yeah, sure," Mason said bitterly. "You think I could take knowing you guys had our horses on top of everything else you've got?"

They were silent. I always thought the Collins couldn't help having a lot of money any more than we could help not having any, but whenever Mason thought about it, it really hacked him off.

Bob understood him, as usual. "You had to sell them, Mason, and you knew Tex would take it hard. You're going to have to quit brooding about it."

"If I hadn't sold them, they would have starved, or we would have. I am not going to quit school and get a full-time job. I'm not even going to get a part-time job till basketball's over with. Basketball is my out. I'll sell anything I have to to make it through this winter. I have *got* to get out of here, man . . . Bob, I've had three different scouts talking to me already. If I can get through this winter, I can go anywhere."

I went to bed, more to get away from that conversation than anything else. If Mason could get a good price for me, I wondered how long it'd take him to make out the bill of sale. Before today I didn't really think he could go off and leave me here by myself if Pop didn't come home. Now I knew. He'd go so quick it'd make my head spin. And Pop—but he had to come home. He always had. He'd just forgot about how long he'd been gone. Anyway, if he knew Mason was gone, he wouldn't leave me here by myself. I couldn't imagine being here by myself.

The house was colder than a frog's tail. I decided to sleep with my clothes on, which I do sometimes anyway, if I don't feel like taking them off. I piled all the quilts on top, instead of saving one to sleep on, and a mattress button bit me all night. We didn't have any sheets. Sheets get torn up real fast, especially if you take a nap with your spurs on.

I woke up at six, just like always, pulled on my boots and jacket, and stumbled out the back door to the garage, where we keep the feed. I noticed we were low on grain, and thought about reminding Mason to get some. Then I went up the well-beaten path to the corral. It wasn't till I got there, wondering why nobody was nickering, that I remembered.

I stood staring at the empty corral, shivering. The horses were gone. Negrito was gone. I'd never find them. It was final.

Once, a couple of years ago, I got into trouble with the law. I borrowed a car that somebody had left the keys in. I took it back, but the police were waiting there for me. They let me sit in a jail cell a couple of hours to think about it before they called Pop.

I remember seeing his face, when he came to get me, almost like he had written me off, like I wasn't his kid anymore. I remember thinking, "This is the worst thing that'll ever happen to me."

And here I was, wrong.

TWO

I sat down at the kitchen table, sore and dazed and half-surprised that the world seemed to be rolling along as usual. Mason came into the kitchen with a bottle of iodine. I looked away while he dabbed some on the cut above my left eye and on the raw knot on my forehead. I couldn't think of anything to say to him.

Then he laid the back of his hand on my forehead and said, "You're running a little bit of a fever—you want to stay home from school today?"

"All right," I said. Then I said, "I'll tell everybody I was sleep-walking and fell down the back steps."

I still wouldn't look at him, but after a second he reached out and patted me on the head like a puppy. I knew we were made up. But things weren't the same.

I knew, in the back of my mind, that Mason had changed a lot during the last couple of years, but now it seemed like I really didn't even know him too well anymore. He used to laugh more and yell less and while he always had been a reserved kind of guy, at least he used to talk to me like I was another human person. Nowadays he was constantly surprising me. I couldn't tell how he was going to act, what he was going to do next.

Like a couple of days later, he announced he wasn't going to the Fair. I was as shocked as if he'd said he was going to skip Christmas.

"How come?" I was sitting in the kitchen, riding a chair backward, watching him make the weekly batch of chili. Mason was really good at chili-making. A funny thing, though, when people tried asking him for the recipe, he never could remember how he did it. I'd watched him a dozen times before, and he never did make it the same way twice.

"Why should I go? It'll just be the same as last year. And the year before that. Rides and crummy sideshows and gyp games and livestock shows . . ."

His voice trailed off. Any time the conversation got near horses, a funny kind of uneasy silence would drop on us like a cloud promising thunder. Since the fight both of us had to work at it to sound like we were carrying on a normal conversation, but now he was thinking the same thing I was. Last year I'd entered Negrito in some of the Western classes at the Fair and won three ribbons, one of them a first place. And Mr. Kencaide, of Kencaide Quarter Horse Ranch, came up to me after and said, "You've done a lot with this little horse, kid. You've done a real good job." Mr. Kencaide is the kind of guy who doesn't go around talking to a lot of people.

Thinking about Negrito made my stomach lurch like I'd dropped off a cliff.

"Anyhow," Mason continued quickly, "it's a waste of money. I didn't like it last year."

"You've been to the Fair every year of your life!" I said.

"So I ought to know what I'm missing, right? I just outgrew it, Tex. That's all there is to it."

He tasted the chili, then added some more peppers. Mason's chili was good and strong.

"Well, I ain't. And I'm going," I said. He didn't say anything.

"I ain't going to outgrow it, either. I'll think the Fair is fun no matter how old I get."

"That," Mason said, "is quite likely."

I breathed a little easier. At least he wasn't going to try to keep me from going.

"Can I take the pickup?" I asked finally.

"Nope. No way. Last time you drove it by yourself you got in that drag race that almost killed you. You wouldn't last ten minutes in the city. Anyway, you don't have a license."

That was pretty corny. We both started driving when we were twelve years old. But he never had let me drive in the city by myself.

"I bet Bob'll drive me and Johnny," I said.

"Probably. Bob doesn't have the sense God gave a goat. Here's three bucks. That's all you're going to get, so don't ask for any more."

I was surprised to get that much. Mason didn't part with money easy. If Pop had been here he probably would have given us ten bucks apiece.

"I wish Pop would come home," I said. Mason looked hacked-off. Just mentioning Pop got him irritated lately. You'd think since he was Pop's favorite kid, he'd be a little easier on him. I'd known Pop liked Mason better ever since I can remember. It was a little hard to figure, since I was the one who acted like him, who wanted to be like him. Pop and Mason weren't anything alike. Pop never took anything too serious. Even that time I got into law trouble, he got over it quicker than I thought he would. Pop was a lot of fun. Mason wasn't exactly Mr. Chuckles.

"Maybe he ain't coming back," Mason said. It seemed like my heart stopped beating for a second, then started off at a gallop.

"What makes you say that?"

"Well, he's never been gone this long before. He's not making any money rodeoing anymore, it's just an excuse to go roaming around with the other good ole boys. I figure one of these days we're going to slip his mind completely."

"You're nuts," I burst out desperately. "He's coming back. You're crazy to talk like that."

"Maybe so," Mason said, but he didn't sound like he meant it.

"Is your hair wet?" he asked suddenly.

"Yeah, I just washed it."

"Again? You washed it yesterday."

"Yeah, well, Jamie said it looked better when I wash it every day."

"Jamie gonna pay the hospital bills when you come down with pneumonia?"

There was a funny kind of quiet. Our mother had died of pneumonia. It was a long time ago.

"I bet Pop's home by Thanksgiving," I said.

Mason said, "You're on."

Sunday was the first day of the Fair and Bob Collins drove us into the city for it. He had a date with him, so me and Johnny had to swear to sit in the back seat and not move and not talk and only breathe enough to keep us alive. Bob had driven us places before.

We lived about twenty miles from the city. That was good in a way, since it was close enough to get to when you got really bored with a small town, and you probably knew what was going on more than somebody in a small town who was a hundred miles away from a city. But it was bad, too, since as soon as you said you were from Garyville, the city kids thought you were a hick. Almost all the kids in our town wore long hair and old jeans and smoked grass and got drunk, just like the city kids, but somehow everybody thought it was more cool when the city kids did that stuff. Especially the city kids.

Me, I liked living in the country and some of the other kids liked it, too. Some of them pretended they did because they couldn't live anywhere else. Then you had the people like Mason, who were itching to get out. I couldn't quite figure out why.

Me and Johnny kept our word and didn't pester Bob a bit on the way to the Fair, in case he decided to turn around and take us back. (We had several things planned for the way home, when he couldn't do much about it.) We did clown around a little bit, but I doubt that he heard us—his girlfriend talked a mile a minute. Bob dropped us off in front of the north entrance to the fairgrounds and told us to meet him there at ten o'clock, then he drove on to the movies or wherever he was going.

It cost a dollar just to get into the Fair, which was a good inspiration for trying to figure out a way to sneak in. I'd managed it the year before, and me and Johnny walked up and down the fences for about half an hour looking for a safe spot to crawl over. Maybe a lot of people had managed it the year before, because this year there were too many cops and guards to even try it.

"Shoot," I said, as we forked over the dollar at the entrance, "that leaves me two bucks. I'll get about two rides out of that."

"Don't worry about it," Johnny said. "I'm loaded."

Johnny never tried to brag about having money—he got five-dollars-a-week allowance, and could usually get some more from either Bob or Jamie, and he'd been saving up for the Fair for a long time. He was real generous, not to show off or anything, it was just that money was something to have a good time with, and if you were his buddy, he wanted you to have a good time, too. I liked that attitude. I never felt bad about letting him pay for anything if I didn't happen to have any money. Mason, man, he wouldn't let you give him the time of the day.

I was sorry Mason was missing the Fair this year, but I was glad to be there with Johnny. Whenever I went with Mason, he just about drove me nuts, because he wanted to plan everything out. He'd walk up and down, and look everything over, and decide exactly which ride was the scariest, and which game he could win at (shooting baskets last year he cleaned up on stuffed animals), and where to buy the cheapest hot dog. We'd be there an hour before we really did anything.

The only kind of plan I ever followed was going through the mile-long livestock barn first, so I could look at the different breeds of horses. But this year I didn't feel like looking at a lot of horses, so after we gawked at the cattle awhile I said, "Let's head on out to the Midway."

Johnny said, "Sure," not asking a question.

Johnny was smarter than most people thought. Once you get a reputation for being scatter-brained, people always think you never have a serious thought in your head, but that isn't always true. I ought to know.

We ran from ride to ride, just hitting the scary ones. Some were fast, and some flipped you upside down, and some whirled you around, and some did all three about a hundred feet in the air. Between rides we ate corn dogs and fried chicken and ice cream and corn on the cob. When we weren't eating or riding we played the games, trying to knock down bottles with baseballs, or pitch nickels in a plate. I won a big red toy poodle rifle-shooting. They always screw up the sights or something on those rifles, but after one round I figure out how to compensate for it, and the rest is easy. I am a very good shot. I hunt ducks mostly.

Johnny won a bear, throwing darts. We felt silly carrying the animals around and gave them to some little kids. Then we watched the preview of the girlie show, but you had to be eighteen to go in. Johnny wanted to go through the freak show, so I stood around outside while he did. I'd been through the freak show before, and last year it depressed me for some reason.

Me and Johnny got kicked off the double Ferris wheel. Even though it was the highest ride at the Fair, giving you a view of the whole city, after some of the other rides it seemed a little tame. So we livened it up a little by rocking back and forth. We were on the top wheel, trying to scare each other, forgetting neither one of us has the sense to be scared. We almost flipped the seat, but then some people down below noticed and the operator brought us down in a hurry, yelling at us before we even came to a stop. He was a mean-looking carnie, and we shot off before he got a chance to do much more than holler, "You kids want to kill yourself, go jump off the Mad Mouse!"

We ran behind a hot dog stand, and leaned against it, laughing.

"Well, I wondered who those two idiots were, and I might have guessed."

I turned around. There was Jamie and some other girl. Jamie had on blue jeans and a blue sweater and a funny blue Fair hat with a big feather in it. She looked real cute.

"We were just having fun," Johnny said. "That's what it's for, isn't it?"

"You moron," Jamie said. Then she said, "This is Marcie, her mother brought us."

I'd seen her friend at school, but hadn't paid much attention to her. Ninth-graders don't mess around with seventh-graders much.

We started down the Midway again, Johnny and Marcie walking together, me talking to Jamie.

"You're almost looking human again," Jamie said. "I'm surprised at how many people believed that sleep-walking story you were giving out."

"Well, I do sleep-walk," I said. People always

seemed to find that fact more interesting than I do. "You didn't tell them any different?"

Jamie shook her head. "Wouldn't do me any good if I did. Mr. Super Cool Mason can do no wrong, as long as we get to the state finals in basketball this year."

Some of the couples we passed had their arms around each other, with one hand in the other's back pocket. That looked like fun, but I didn't think Jamie would go for it.

"You guys been to the fortune teller yet?" she asked.

"Naw, we wouldn't waste money on junk like that," Johnny answered.

"She really is good," Jamie said. "Come on, let's go. I want to hear what she tells you guys."

We followed along. Jamie had a way of making you do what she said. Anyway, if the fortune teller was any good, maybe I could find out where Negrito was. We stopped outside the booth.

"No way," Johnny said. "I'm not going to throw away a whole dollar."

"Well, hell, I'll pay for it." Jamie slung her purse off her shoulder.

"Cole's going to cream you if he hears you cussing like that anymore."

"He's not here to hear it, is he? Go on in."

Johnny took the dollar, winking at me. One of his favorite tricks was to see how much money he could worm out of his family. He was good at it, but then, he wasn't related to Mason.

"Maybe you guys could ride home with Bob and us," I said.

"No thanks," Jamie said. "I'm around them enough."

"You know, Jamie," Marcie giggled, "your brother is really kind of cute."

"You should see Charlie," Jamie said shortly. I got the definite feeling she didn't like other girls being interested in her brothers.

Johnny came out of the small tent. "Did she say you were going or staying?" Jamie asked him.

"Staying," Johnny said.

"What are you talking about?"

"You'll see." Jamie started to get out another dollar.

"I'll pay for it," I said. It was my last dollar, and one Johnny had loaned me at that, and I sort of wanted to save it to do the shooting gallery again. Jamie liked stuffed animals. But I wouldn't feel right letting her pay for anything.

I went inside the tent. It was kind of cramped, and so dark after the neon glare outside that I had a hard time seeing at first. With the curtains pulled shut, all the Fair noises seemed muffled and far away. The place sort of gave me the creeps.

"Sit down please."

My eyes had adjusted enough to see a lady sitting behind a table. She was younger than I'd thought she'd be, dressed up like a gypsy. You could tell it was a costume, though, not something she'd wear all the time. I sat down. I'd never been in that kind of a set-up, and I wasn't sure what to do.

"Cross my palm with silver." She held out her hand. Now if Johnny had been in there with me, the whole thing would have been funny. But alone, I didn't feel like laughing. The hairs on the back of my neck were tickling me. I had that happen before once, when I seen a ghost.

And now even the little hairs up and down my backbone were standing up, tickling me. This place was creepy.

"You mean pay you?" I asked, after a minute. It'd look funny to Johnny and them if I rushed out without getting my fortune told. She nodded, so I handed her the dollar.

"When is your birthday?"

"October twenty-second. I'll be fifteen," I added. She looked at some kind of map spread out on the table. Then she said, "Let me see your palm."

I held out my hand. She took it in a firm hold and looked at my palm for a minute or two. Then she said, "Your far past: You are a fourth-generation cowboy. Your near past: violence and sorrow. Your next year: change. My best advice: Don't change. Your future: There are people who go, people who stay. You will stay."

She dropped my hand. "You may think to yourself one yes or no question."

That was what I was waiting for. I thought "Will I get Negrito back?"

She was quiet, then said, "I'm sorry, the answer is no."

Up till then she'd been using a fakey, gypsy-type voice, and to hear her turn human on me was the scariest part of the whole thing. I got up, glad to get out of there.

Johnny and Jamie and Marcie were waiting for me. It was hard to get back into the mood of the Fair. I was still thinking about Negrito.

"What did she say? Going or staying?" Jamie asked.

"Staying, I reckon," I said, "whatever that means."

"I think she's a fake," Jamie said suddenly.

"Now's a fine time to decide that. After wasting all that money," Johnny said.

"Well, it wasn't your money so shut up. I think she's a fake. What else did she tell you, Tex?"

"Said I was a fourth-generation cowboy, and I'd had violence in my near past."

"Great. You walk in there wearing boots and a cowboy hat, with the remains of a fist fight all over your face, and she sees you're a cowboy with a violent past. Real powers, all right."

I'd forgot my face hadn't quite healed up.

"Anyway," Jamie went on, "for you to be a fourth-generation cowboy, your father and grandfather and great-grandfather would have to be cowboys. And you told us your grandpa had been a preacher."

That was right. Pop used to tell us about the wild stuff he did when he was a kid, then say,

"Well, I was the preacher's kid, so what could you expect?"

"What did she tell you?" I asked Jamie. I got the feeling that Jamie was a person who was going.

"That I'd be married three times, and I know I couldn't stand it once."

"Good," said Johnny, "that'll save three guys a lot of grief and misery."

I wanted to ride something real quick, to get back into the mood of the Fair. The Zipper was one of the scariest rides they had, so we rode it next, Johnny buying the tickets for all of us. Jamie sat with me, and Marcie rode with Johnny. We could hear them both hollering, while Jamie'd gasp, "I don't see what's so scary about this." It flipped us upside down and went straight up in the air and came straight for the ground.

"It ain't scary, it's just fun," I told her.

"You're crazy, Tex."

We spent the rest of the evening paired off. It was almost as good as having real girlfriends.

I thought about cutting through the livestock barns on the way out, for one quick look at the horses, but didn't. I'd got back to being happy, which is the right way to leave the Fair, and there wasn't any sense in doing something you knew would make you sad.

I was real sorry when ten o'clock came and we had to split up—Jamie and Marcie going to the south entrance to meet Marcie's mother, Johnny and me heading for the north to meet

Bob. I could be wrong, but I think they were sorry we had to split up, too.

We were a little late meeting Bob. He had parked down a side street to wait for us, and we tried sneaking up on him and his date to see if we could catch them making out, but they were just talking.

"You want to go by and see Charlie before we go home?" Bob asked, when we got in the car. Charlie was the oldest Collins kid. He went to med school and lived in an apartment in the city.

"Sure," Johnny said, "we haven't seen Charlie in a long time."

"Mason mind you being out late?" Bob asked me.

I shook my head. The only thing Mason minded me doing was skipping school. I could come in at five in the morning, just as long as I was up at six to get ready for school.

"How about you?" I asked Johnny. "Won't Cole get upset if you're out late?"

Johnny grinned. "Not as long as I'm with Bob. Bobby can do no wrong."

For some reason that remark set Bob's girlfriend into a fit of giggles. Bob just kept his eyes on the road.

Walking into Charlie's place was like still being at the Fair, only without the rides. There were a whole bunch of people there, running in and out. As soon as we got there Charlie introduced us around as his family: Bob was his younger twin, he said, and since they looked an awful lot alike, people seemed to accept it. But he also told them me and Johnny were the second set of twins in the family, and our names were Mutt and Jeff, and he told people Bob's girlfriend was their aunt. And some of the people there were in a condition to accept that, too.

Charlie Collins was one of the blonde Collins, like Cole. Everybody liked Charlie. He knew it and wasn't afraid to let you know he knew it. And everybody liked him, anyway. He wasn't as big as Cole, but nearly, and rugged-looking. Like a whiskey ad in a magazine. Clothes that would look silly on anybody else looked fine on him. Jamie had always called him Jet Set Charlie, and that was what he looked like.

Right after we got there he fixed us a drink. I thought it was 7-Up till I took a big gulp and choked on it.

"Charlie . . ." I heard Bob protesting, and Charlie said, "Come on, Saint Robert, I don't get to see you guys that much anymore."

And Bob ended up having a drink, too.

I was pretty thirsty after all that running around at the Fair, and I gulped down my first drink, and the second one, too. It tasted really good, like pop with a zing to it. The only thing I'd ever tried drinking before was beer, and I never could get a taste for it. Beer tasted awful sour to me. This stuff was sweet. I was on my third one when I noticed things were looking extra bright and sharp and I had to talk louder because my hearing was getting funny.

I'd never been drunk before. I know that's

hard to believe, me being so close to fifteen years old, but it was the truth. Pop never had been much for booze, partly, I think, because his prison stretch had something to do with bootlegging, but mostly because he didn't need it. Pop always had a good time.

Poor Bob kept trying to stop the flow of alcohol and kept ending up with another drink himself. Johnny got out a bunch of medical books and laughed his head off at the gory pictures. I sat next to Bob's girlfriend and talked a blue streak. I talk a lot anyway. I think I talked a lot about Jamie, but she didn't pay much attention to me. She never took her eyes off Charlie, who laughed and told funny stories and got out his guitar and made up dirty songs (he really could sing) and met people at the door and waved good-bye and passed the drinks around one more time. Charlie was an extremely likeable person.

"Charlie," I said to him seriously, trying to get my eyes focused, "there are people who go places, and people who stay, and I think you're going."

"I've been going all my life, cowboy," he grinned at me. "And I love it."

I thought that was nice, and wondered how I was going to like staying.

Johnny laid down on the floor and flapped his arms like he was swimming. It made perfect sense at the time and I thought about joining him.

Charlie was trying to get us to stay all night.

"I'll call Cole and tell him you've got car trouble."

"I am fine," Bob told him, "I am perfectly fine. I have never had a driving accent, accent . . . I mean wreck. Come on, Jonathan."

Bob drunk and trying to act dignified set me and Johnny off on hysterics. But after all he was the driver and therefore boss, so I tried to pull Johnny up on his feet and ended up rolling around on the floor with him. Charlie finally got us standing upright.

"Bobby, you sure you can drive, kid?"

"Certainly."

Bob's girlfriend belted down her fourth or fifth drink, got up, and said "God, you're gorgeous!" to Charlie. It was the first thing she'd said since we got there. Charlie kissed her on the forehead, hugged Johnny and Bob one more time, shook hands with me, and made sure we didn't fall down the stairs on top of some more people who were coming up.

"Visiting him is like getting hit by a tornado," Johnny mumbled. Both of us wanted to lay down in the back seat, and we got into a shoving match over it. We finally ended up flopped together like puppies in a litter.

The next thing I knew I was home, puking off the front steps into the bushes.

"Mace," I gasped, for some reason sure he was there with me, even though I didn't remember seeing him, "I think I'm sick."

"I think you're drunk." Sure enough, his voice was right there next to me in the dark.

"Robert was drunk, and Johnny was drunk, so I imagine you're drunk, too, seeing how you don't have any tendencies toward car sickness."

Mace pulled me back up before I did a headlong flip over the porch railing.

"Are you mad?"

"Oh, no, I'm thrilled to death. Now get to bed."

On the way there I could tell I was going to throw up again. I detoured into the john, but ended up puking in the tub instead. It was easy to see what kind of junk I'd been eating at the Fair, and it made me sicker. I thought my insides were going to come up. I broke out in a cold sweat all over. Even the inside of my mouth was sweating. Mason handed me a towel and kept me from falling down when I tried to stand up again. When I made it to bed, I curled up and shivered like I was freezing, but I was really hot. The whole room was turning around and around and there were minutes when I'd think I was still on some ride at the Fair, even to hearing the music and noises. I'd heard of people passing out from drinking too much, and figured maybe I would, too, but I think I went to sleep first.

"Tex, wake up. Come on, it's okay, you can wake up."

Slowly the room took shape around me. It was still dark. I was at the front door, hanging on to the doorknob. Moonlight was shining in the windows, making the bare floorboards gleam like water.

"Am I asleep or awake?" I whispered. I couldn't tell. I was so scared I couldn't move.

"You're awake now. Come on. You know you're awake." Mason was right there beside me, talking in an easy voice, like he was quieting a spooked horse. He never gets mad at me when I sleep-walk.

He hadn't been touching me, but now he reached over and gently pried my hand off the doorknob. "Come on, you're still drunk, that's all."

"Was I hollerin'?" I had a vague memory of somebody yelling.

"A little bit, not much, come on back to bed."

I let him steer me back to the bedroom. Waking up from that nightmare always leaves me so scared I'm almost paralyzed.

"You ever remember what you're dreaming when you do that?" Mason wasn't asleep like I thought. "I always find you at the front door. Lots of times you don't even wake up, I just lead you back here and you go back to bed."

"Yeah, I remember," I whispered. I was still shaky. "You know that fight Mom and Pop had, just before she died, when she ended up walking out in the snow? I dream about that."

"Geez, Texas, you couldn't remember that! You weren't three years old."

"Well, I remember it, anyway. Sort of. The yelling, everybody seemed so tall, and when she walked out I was trying to stop her, go after her, but I couldn't reach the doorknob."

"I can't believe you remember all that."

I was quiet. I did remember it. Not real clearly, but I did remember some. But it always seemed to me that she died right after that, that she never came back, even though I knew she was home a couple of days before going to the hospital, and she lived for a couple of more days there.

I remember crying a lot at that time, too.

"Sorry I woke you up," I said.

"In the daytime you aren't afraid of anything," Mason said.

That wasn't completely true. There were people who go places and people who stay, and Mason was going. I was afraid of that.

THREE

L

Mason was yelling "Okay, okay, I'm comin', I'm comin'!" and as I woke up I realized somebody was knocking on the door.

My head was pounding. It was like the insides had swollen so much they were pushing on my skull. My eyes hurt real bad. I never noticed the bedroom light being so bright and glaring before. I rolled over to push my face into my pillow, and my stomach couldn't keep up with the rest of my body. For a second I thought I was going to barf.

I must have the flu, I thought vaguely. Man, I was sick.

The bedroom door was wide open and the front door wasn't much on the other side of it, and I heard Mason say, "What do you want?"

He has a blunt, rough way of talking that irritates a lot of people, but he sounded even sharper than usual.

"Bob and Johnny came home drunk last night, and I'd like to know why."

It was Cole Collins! I got up quick, with some crazy idea of hiding in the closet, but the room spun around so wild I ended up on the floor instead. I slid under the bed, and lay there gripping the floorboards with my fingernails. The cold floor felt real good against my hot face. Cole Collins!

"They were drinking, I imagine." Mason sounded real calm.

"I know Bob. He's not the kind of kid to do something like that, and he certainly would never let Johnny drink." Cole sounded impatient.

"Well," Mason said, measuring out his words like he was scared he was going to drop one, "if Bob wouldn't drink and wouldn't let Johnny, it sure is strange they came home drunk, isn't it?"

Dear God, I prayed, don't let them get into a fight. Cole Collins was six-four and must have weighed two hundred pounds.

"So somebody got them drunk. And I'd like to know who."

"It wasn't me." Mason bit off each word. He sounded like as far as he was concerned, the conversation was over.

Then Cole said, "Where's your brother?"

I froze. I shut my eyes so tight I saw sparks.

Tell him I'm at school. Tell him I left home. Say that I . . .

"It wasn't him either. Why don't you ask Robert who it was?"

There was a silence. Then Cole said, "Bob says he's to blame. He takes full responsibility. He isn't a good liar, though, even to protect someone else . . . I'd like to talk to Tex, if you don't mind."

I had an image of him searching the room, dragging me out from under the bed . . . if Mason could knock my teeth loose, Cole Collins could knock my head right off my shoulders.

"I mind all right," Mason was saying. "I mind a lot, as a matter of fact. Don't you have enough kids of your own to hassle?"

My eyes flew wide open. Now he's done it! I started inching my way out, figuring if they got into it I could help Mason.

Cole was quiet, like he was so mad he couldn't talk. Then he said, "I'd appreciate it if in the future you didn't associate with my kids."

"Tell it to them!" Mason said, and slammed the door so hard the whole house rattled. The sound split through my head like a bolt of lightning.

Mason was swearing a blue streak. I had decided to crawl out from under the bed, but on second thought stayed where I was. Mason came stomping in. "I'd like to know who he thinks he is—associate with his kids, hell! I reckon he thinks we'll corrupt the little darlin's.

So Bob's protecting somebody, huh? I could sure tell you who, you—"

He broke off suddenly. "Tex?"

I scooted out from under the bed. I was grimy with dust balls and cobwebs. My stomach started churning around again, and when I sat up funny black lines kept floating across my vision.

"Lose something?" Mason asked, sarcastic. "Well, get up and clean the tub out. You'll sure have to take a bath before school."

"I c-c-can't . . ." I broke off, sneezing from the dust all over my face. Every sneeze felt like it was going to pop my head wide open. I held my head together with my hands. "I can't go to school," I finally finished. "I got the flu."

"You got a hangover and you're going anyway."

"This is a hangover?" I asked, amazed. "But Mason, I'm really sick!"

I thought about the times Roger Genet came to class drunk or hungover and he groaned about how miserable he was, and everybody'd laugh at him. Shoot, I wouldn't laugh at him no more. What I couldn't figure was, if drinking made you sick, why anybody'd want to do it?

I got up queasily, went to the john, took one look in the tub, and got a case of the heaves. I leaned on the sink, trying to get steady. I got a look at myself in the shaving mirror. All the blood had drained out of my face till my tan looked like a layer of brown paint over white.

"All right," Mason said, pulling on his boots.

"You can stay home. Just get that mess cleaned up later."

He gave a short, sudden, mean laugh.

"What you thinkin' about?" I asked, crawling past him to get under the quilts.

"I was thinking about going over to the Collins' and demanding to know who got you drunk last night."

"Hey don't do that," I said, alarmed. "It was Charlie."

"Oh, I knew that. But if Robert wasn't going to squeal, I wasn't either."

I was shivering. "There any cure for a hangover?"

"Aspirins, but we're out. A Bloody Mary's supposed to help."

"What's that?"

"Vodka and tomato juice."

"Vodka!" I groaned. "No, thanks."

I spent a lot of that day in bed, sleeping on and off. When I did finally get up, it was like all the energy had been drained out of me. I cleaned up the tub and felt so bad I did the dishes just because I couldn't think of anything else to do. Then I went outside and sat in our tire swing, staring at the empty horse pen, thinking about Negrito. Pop had given me Negrito three years ago Christmas.

It was Monday. I'd miss vocabulary day in English. Well, Miss Carlson would be glad to know I added a new word to my vocabulary, anyway. Depressed. Man, I never knew what the word meant before. I could see where they got it. It was like everything was pressing down on you like a dead weight. I'd been sad before, not much, but some. But man, depressed! It was worse. Like you couldn't see the use of anything, everything was just hopeless. I sat out there and twirled around in the swing, watching the dead leaves blow by. Summer over, Negrito gone, Pop gone for so long I'd just about given up on him, Mason planning on hightailing it off as soon as he could, maybe the Collins couldn't be friends with us anymore. It was like the future was a foggy pit and I was standing on the edge, trying to see the bottom, knowing any minute something was going to shove me in. It was a real uncomfortable feeling.

I sat out there till Mason got back from basketball practice. He brought a carton of ice cream with him, which was lucky, since one whiff of the chili we were having for supper was enough to give me the queasies again.

I couldn't even eat much ice cream, watching Mason wolf down the chili. He eats like a horse.

I spent the evening cleaning all the guns in the house; my .22, Mason's 20-gauge, Pop's old double-barrel 12-gauge. It was something easy to do, that wouldn't put too much strain on my aching head. Mason was studying as usual.

Finally I got a little hungry and got out the ice cream again. There wasn't much left. Mason had really pigged it up.

"What the hell is all that honking about?" He looked up from his book.

I listened. Usually it's so quiet around here you can hear a car coming a mile off. This one was honking the horn full blast.

It got louder and louder, then the car pulled into our driveway on one long last blast.

"Hey, maybe it's—" I jumped up and ran to the door.

"Mason!" a voice hollered. "Tex!"

"It's Lem!" I said. For a second I was sick that it wasn't Pop—then forgot it because I was real glad to see Lem.

He covered the yard in about two strides, and I jumped him in a flying tackle from the porch. He staggered, but didn't fall down, swung me around and let go, throwing me halfway across the yard. He jumped the porch steps to give Mason a bear hug. Mason was laughing. It was the first time he'd really laughed in a long time.

Lem Peters was a friend from way back. We'd known him longer than we'd known the Collins. In fact one of the first things I can remember in my life is hunting snakes under Lem's house. He and Mason had been best friends for years, then they kind of split up until they weren't best friends anymore, but still pretty good friends.

Lem was real good with horses. When we were little kids, all three of us would pile on Lem's white mare and go for miles across the countryside. Lem was someone who had always been there, so it had seemed strange when he got married and moved away last year. It was sort of the first time I realized that things weren't going to stay the same all my life.

Mason was real glad to see him. He glanced at me sideways like a wicked colt and said, "Where's the wife?"

"The wife?" Lem answered innocently, trooping in the house with us, dropping down on the sofa, just missing sitting on my ice cream. "Well, I reckon she's with another man."

Mason and me looked at each other. I had heard enough of Mason's dire predictions about Lem's marriage to get a little worried; but Lem didn't look mad or anything. He looked more like he was about to bust. And sure enough, in a minute he did, with, "But seein' how he's only a couple of hours old, I don't think much will come of it, even if she is crazy about him."

Mason got it before I did. He jumped up and yelled, "You're kidding!"

Lem just beamed at him. I never saw a sober person look so silly. Suddenly it dawned on me.

"You had your baby!" I said.

"Nope. Connie had my baby." Lem looked like a cat with kittens. If you know cats you know what I mean. Everytime a cat has a batch of kittens, she thinks it's the first and only set of kittens in the world. I've had several of them tell me so.

All three of us were dancing around the room,

laughing our heads off. Finally we dropped to the floor.

"A boy, huh?" I said.

"Yep. Little Lem."

"Don't call him that," Mason said. He was still laughing, but there was a streak of serious in his voice. "You don't want another Lem Peters."

"Well, to tell the truth, he does look more like Connie, kind of blonde and pug-nosed. Think maybe we ought to call him little Connie instead?"

Mason just shook his head. He didn't much like being named after Pop. I wouldn't have minded it. Mace reached out to slap Lem on the back. "Tell your folks yet?"

Lem's craggy face darkened. "Nope. They ain't gonna find it out from me, either. All I heard when me and Connie got married was, 'Don't bring a squalling brat home for us to raise.' Well, they don't need to worry about that."

We were quiet for a minute. Connie's family hadn't been thrilled with the marriage, either.

"How's it going in the city?" I asked. It was hard to picture Lem in the city. He was long-legged, gawky, worse than me about bumping into things and knocking things over. There would be a lot more things to bump into and knock over in the city.

"All right. I got a job at a gas station. It don't pay much, though. Connie had a job for a spell, but it'll be a long time before she can go to work again. Anyway, I don't want her working."

"Hey." Lem brightened up. "I ain't told the Collins yet. Let's go up there."

"This time of night?" Mason said. "You think Cole Collins is gonna break out the champagne or something?"

"Shoot, Cole don't have to know anything about it. We've been in that house when he didn't know anything about it. Let's go."

"Mason . . ." I said, thinking about what had happened that morning. "Maybe . . ."

"Can this be fearless Tex McCormick, the terror of the town?" Mason mocked me. Lem was looking at me funny, too. He'd always let me tag along, from the very first, because I'd never been scared to do anything he or Mason would do.

"Okay," I said. Cole Collins scared me a little. Not 'cause he was big or rich or vice-president of a company. He just plain didn't act like I thought fathers were supposed to act.

We started off across the field. The cold night air made me feel better than I had all day. I love to go roaming around at night.

"You know," Lem said as we walked along, "I reckon it was a real shock to ole Cole to find out there was people like you and me livin' out here. I think he moved here to get his kids away from all those bad influences in the city."

"I don't care why he moved here," Mason said shortly. "And as far as bad influences go, last night it was one of his do-no-evil heirs that got Tex drunk for the first time in his life."

"Geez. Tex, what were you waitin' on?" Lem asked me. "Your first gray hair?"

"I'll go on ahead and let the dog know we're coming," I said hastily. Mason hadn't bawled me out about last night yet, and I didn't feel like getting it in front of Lem.

The Collins dog gave a short bark as I came running up, but then he recognized me and started wagging his tail. Fortunately, he barked at everything that moved, squirrels, owls, a piece of paper, so the Collins didn't pay too much attention to him anymore. But he would have set up a real racket for three people coming toward the house in the dark, so it was a good thing I went on ahead.

Johnny's room was on the first floor, just off the kitchen. In the summer it was easy for him to get out, or us to get in, but since it was getting colder and the windows were shut, we had to risk making some racket before he woke up and let us in.

"What's up?" he whispered. "Tex! Hey, what are you doing here, Lem?"

We climbed in the window one at a time. We were laughing so hard it wasn't easy keeping quiet.

"What's up?" Johnny kept whispering, and we couldn't get our breath to tell him. Across the room Bob sat up in bed and said sleepily, "What the . . . ?"

We piled onto Johnny's bed, while he sat on the floor and leaned back against Bob's.

"Guess what I got?" Lem said finally.

"Measles," Johnny said. "Ticks. A job—" He broke off suddenly, trying to hush us up. "Be quiet you guys. I'm grounded for two weeks already. I don't want to try for a life sentence."

"I'm not afraid of Cole," Mason said. If he really wasn't, he was the only person in the room; at least the rest of us quieted down.

"A baby," Lem said. Johnny just looked at him blankly.

"A baby what?"

I shoved Lem off the bed, and he dragged me with him, and all of us were biting our sleeves to keep from howling.

"I wish I could tell Blackie and Jamie," Lem said.

"Blackie's moved out, he's in San Francisco now, painting," Johnny said. "And I'm not about to go get Jamie. Cole's still up. I'm in enough trouble."

"I'll go get her," I said.

Mason said, "More guts than brains, as usual." But he didn't really try to stop me.

I had to go past the kitchen to the stairway, which was just off the living room. I crawled down the hall on my hands and knees. I've been hunting since I was seven years old, so I know how to get some place without being seen. I peeked around the corner of the main hall into the living room. Cole was sitting at his big rolltop desk, writing, his wife, Mona, curled up in an overstuffed chair, reading.

I started crawling up the stairs. Man, I was putting something over on Cole Collins.

The first bedroom off the stairs was the one Charlie and Blackie had shared before they left home. Across the hall was Jamie's room. Cole and Mona's room was way down at the end of the hall. Like Lem said, we'd been in that house before without Cole knowing it.

I pushed Jamie's door open slowly, trying not to make any noise. Jamie lay facedown on the bed, hugging a pillow. She looked so soft, laying there, curled up like a kitten. She'd probably feel like a kitten to touch. I had a funny feeling in my stomach, looking at her. It was like when you get out of a swimming pool and lay on hot cement and your belly caves in. I reached over to wake her up.

"Hey!" I whispered. I didn't have time to touch her. Quick as lightning, she whipped a water pistol out from under her pillow and squirted me right in the face with it.

"Hey!" I said again, backing up into the door and slamming it shut.

"I thought you were Johnny!" she whispered. I was wiping my face with my sleeve. "Well, I ain't."

Then I froze. Somebody was coming up the stairs. And it was too heavy to be Mona. Jamie pointed frantically to the closet. I scooted in just as Cole opened the door. I stood there trying not to breathe or jingle a hanger. My heart was pounding so loud I thought sure he'd hear it.

"Jamie, did I hear something?"

"Oh." She sounded so sleepy I could almost believe she'd just woke up. "I guess the wind blew the door shut."

"I've told you not to leave your window open in this weather."

"I can't stand sleeping with closed windows."

"You won't be able to sit if you don't."

Cole walked right by me, over to the window to pull it down. If he glanced into the closet he'd see me. I hadn't had time to slide the door shut. I didn't even want to wonder what he'd do.

"Leave it shut, you hear?"

"I'm not deaf," Jamie muttered.

"What?"

"Yessir," she said louder. Cole paused, like he was going to say something else, then sighed and went down the hall.

I waited a minute, then came out.

"I thought you were Johnny, playing ghost or something," she whispered.

"Well, that's as close to being a ghost as I want to get," I said. "Boy howdy, you are always prepared, ain't you?"

"You'd better believe it. What's going on?"

"Oh, yeah, Lem's here, down in Johnny's room."

"Lem? Oh, I guess he's had his blessed event, huh?"

"How'd you know?"

"Well, it wasn't a big secret he was expecting one, was it? I get A's in math, I can count to nine."

"Come on," I said, "let's go see him." I was still a little shaky from that close call. If I had to get caught in that house, I figured I'd be much better off getting caught in Johnny's room than Jamie's.

We got back down the stairs okay. Johnny and Bob and Mason were sitting on one bed, while Lem was showing them how he'd paced the floor in the waiting room, smoking cigarettes, just like you see expectant fathers in cartoons. Everybody was laughing and trying not to make any noise.

"Get in a rain storm on the way?" Mason asked me.

"Trigger happy Jamie strikes again," Johnny said. "Be thankful it wasn't loaded with ink."

And when Bob started to open his mouth, Jamie cut him off by saying, "It could have been my BB gun instead of my water pistol. So skip the lecture, Bobby."

"Maybe you better go back and try getting out of the other side of bed," Bob said. I didn't want him fussing at Jamie, so I said, "Cole almost caught me in Jamie's room."

"Man, can you imagine?" Jamie said. "Instant death."

"Well, you are a little young to be keeping men in the closet," Johnny said.

"If I had my way, that's where they'd all be." She turned to Lem. "So what was it?"

"A boy," Lem said proudly.

"Yeah?" said Jamie. "Well, don't worry. If it'd been a girl you could have just tried again until you got what you wanted."

"Geez, don't get her started on her women's lib stuff," Johnny groaned.

"I ought to know something about discrimination—look who has a motorcycle and who doesn't," Jamie said.

"I'm glad Connie doesn't mess around with that stuff," Lem remarked. "I think women ought to stay home and take care of their husbands and kids."

Jamie's eyes widened, then narrowed down like a fighting cat's. Johnny gave me a look that said, "Oh, brother, now he's done it!"

"If Connie can figure out which end to diaper, she'll be doing better than most people expected, so I'm not surprised she doesn't 'mess around with that stuff.' "

Lem, like Mason, was used to treating Jamie like a little kid. I was the one who noticed she wasn't a little kid anymore.

"Jamie, you don't know nothin' about being a wife and mother, so just hush."

Jamie started to say something, caught a warning glance from Bob, and stopped for a second. Then she said, "Well, good luck with the prince and heir, Lem. You're going to need it. Personally, I don't think you two could raise a cat."

Everybody sat there, the fun drained out of them. Nobody had wanted to think about the serious part of having a little kid. Lem looked tired all of a sudden, older, like he'd been reminded of things he'd been trying hard to forget.

"You sure know how to cheer people up," Mason said to Jamie.

"Lay off," I said sharply. Mason gave me a funny look.

Jamie jumped up. "I was only telling the truth and you know it!"

She stomped out and slammed the door, on purpose.

"Great," said Johnny.

We didn't need any more encouragement for leaving, but Cole's voice on the stairway, talking to Jamie, gave us a good motive for speeding it up. We climbed out the window in a much bigger hurry than we climbed in it.

"Here comes Cole!" Johnny whispered. "Seeya." He almost closed the window down on Lem's leg. We flattened ourselves up against the side of the house. If Cole looked out the window he'd better not see anybody.

We walked back across the field in silence. Mason and Lem were covering a lot of ground, the way long-legged people usually do, but I could have run circles around them. Night air affects me like that. I didn't, because it seemed like it would be disrespectful to Lem, he looked so sad and tired.

He stopped on the porch instead of coming in with us. "I better get back. I got to get up early and go to work."

"Tell Little Lem hey for us, huh?" I said.

"Sure." Lem grinned a little. "I forgot to hand out my cigars."

He took a handful of joints out of his shirt pocket. "Made up special, too. A guy I know at work gets it for me now. Real good stuff."

I took one. It was almost as big as a cigar. Mason shook his head. "I'm in training."

"Oh, yeah, basketball star. You'll have to teach Lem Jr. the fast break."

"Sure," Mason said. Everybody was trying to sound cheerful, but it wasn't real like it was before.

"I'll see you guys real soon. Come by when you get to the city."

We watched Lem drive off. Mason went on in the house, but I stayed on the porch and thought about smoking the joint.

"If you smoke that thing don't come hollerin' to me in the middle of the night, dreaming something crazy," called Mason from inside.

I thought about last night and put the joint in my pocket. Grass never did do much for me anyway. One of the few times I ever *fell* off a horse (not *thrown* off) I was high. A spunky horse on a cold morning was the best kind of high for me.

"You know," I said, coming into the house, "I never notice how beat up our stuff is till I been at the Collins' for a visit."

The stuffing was coming out of a sofa cushion, and I hadn't even seen it before.

"I notice it all the time," Mace said dryly. He

sat in the arm chair, biting at his lower lip and rubbing his stomach. That was a habit he'd just come up with recently.

"I wonder why Jamie was so hateful. I guess Lem can have a baby if he wants to."

"Apparently he can even if he doesn't want to." Mason paused, then he said, "She's right. Jamie's pretty smart for a spoiled brat."

"But you sounded like you was happy for him."

"Happy for him? Scrounging around, beating the bushes for money, married to an empty-headed bottle of peroxide. Jamie was right. They aren't fit to raise a cat. Sure, I'm tickled pink for him."

"What is with you, Mace? You always sound mad lately," I said. "Lem ain't mad about it."

"He's scared silly."

"He looked happy to me, till Jamie started in on him," I said stubbornly.

Mason stood up and started to stretch, stopping suddenly. "Tex, you are not stupid, and you're not all that ignorant. But how anybody as simple-minded as you are has managed to survive for fourteen years is beyond me."

"Well, I had a wonderful smart sweet brother lookin' out for me," I said. I'm not sarcastic by nature, but I reckon you can learn anything if you're around it long enough.

FOUR

"What a comedown." Jamie hung over the back of our seat and mocked us. "Having to ride the bus like other mere mortals."

"A whole week, too," Johnny grumbled. "No cycle for a whole week."

"At least Cole didn't haul it off to the junkyard like he threatened to at first. Bob's lucky, Mason can give him a ride."

"Yeah," I said, "if he wants to get there an hour early and stay two hours late. Mace is getting to be a fanatic about basketball."

Johnny sighed. Having a cycle at our school, where not too many people had them, really had done a lot for his ego. "You bring any snakes or frogs today, Tex?" he asked.

"Nope. Everything is hibernating."

"There's nothing to do. We already liberated

the ant colony and it's too cold for a water-gun war."

"Aren't you in enough trouble?" Jamie asked.

Johnny shrugged. "I don't know. I'm just so sick of Cole bossing me around. I feel like really giving him something to worry about."

Jamie settled back in her seat. "You can always go out and get drunk again. That worked real well."

Johnny didn't look too inclined to take that suggestion. I didn't tell him my idea for finishing my art project. I could tell it would just make him glum for not being able to do it himself.

I usually like art class. Most of the time I draw horses, which I happen to be real good at, or paint landscapes, which I can do well enough. Mostly I get C's in art, though, because Mrs. Germanie counts off for behavior.

The last three weeks had been driving me crazy. We were supposed to be making a free-form sculpture by gluing toothpicks together. Some of the kids really got off on that. I'm not much on free-form sculpture. As a matter of fact the whole project bored me out of my mind, until the last few days, when I thought of how I was going to finish it. I went nuts gluing toothpicks. Used lots of glue. Now I had a huge sticky mess about three feet high.

"Today is the final day of this project," Mrs. Germanie announced.

"Thank the Lord," I whispered, and the girl sitting at the next table giggled.

"I'll be passing by each desk during the hour to give you your final grade on it. So have it ready."

I was ready. When she got to the table just before mine, I set a match to my pile of toothpicks and jumped back as the flames shot up. "Grade it! Quick, grade it!" I hollered. I thought I'd crack up at the look on her face. Then she just stood there, watched it, wrote a grade down in her little book, and said, "Texas, make sure it's out before you go see Mrs. Johnson."

Mrs. Johnson was the vice-principal and guidance counselor, and she was also the person to see when you got into trouble.

Mrs. Johnson wasn't real surprised to see me. I'd been to her office before.

I got the usual lecture and a swat with the board of education (that's what Mrs. Johnson called it), and finally she said, "Let's not see you in here again, Texas."

"Not this year, anyway," I said. "Maybe not anymore this month."

"Try to stay away until next week, anyway."

I nodded and waved as I left. Mrs. Johnson was real nice. I know that sounds funny coming from me, since I was the next-to-the-most swatted kid in the school. Roger Genet was the most, but he was mean. I always knew when I was doing something I could get swatted for. It was never any big surprise. See, Mrs. Johnson might swat me once in a while, but she always asked me how we were getting along, if we'd heard from Pop, and once she told me about some trouble Mason had been in when he was my age that I had never heard anything about from anyone else. She'd always say hi to me in the halls, and when I came to school after that fight me and Mason had, she spent five minutes trying to find out what happened (I wouldn't tell) and then another five telling me the best thing to do for a black eye. If you get the feeling somebody cares about what happens to you, then you don't mind it if they swat you once in a while.

It just wasn't my day. Miss Carlson asked me to stay after English class.

"Tex, you can't do *two* book reports on *Smoky the Cowhorse*."

"But I read it two times." I didn't tell her I'd read it two times last year, too. *Smoky the Cowhorse* was my favorite book. It had some real good pictures in it, too.

"Why don't you read another book by the same author?"

"You mean the same person might have wrote something else?" I don't know why I never thought of that. I guess I figured writing one book ought to last somebody a lifetime. I don't know how they sat still that long. A two-page book report wore me out.

"Written." Miss Carlson was always correcting our grammar. She was pretty young for a teacher. I don't know, once they get past twenty it's hard to tell how old they are, unless they're really old.

"Yes, Will James wrote several books. He did

his own illustrations, too. I think you'd enjoy *Lone Cowboy*. Tex, have you ever thought about writing poetry?"

Poetry! Holy cow! I glanced around to make sure nobody had heard her say that. What had I ever done to make Miss Carlson think I ought to write poetry?

"No," I said finally.

"You might be good at it. Until you hand in another book report, I'll have to put down an incomplete for your grade."

"Okay. I'll do you another one," I said hastily. I was anxious to get out of there, not because of her, or what she was talking about—even though the poetry thing shook me. I really meant to get that book out of the library. But I was on my way to gym, and I didn't want to be late.

I hated gym. In some classes the teacher is mean to everybody, I can take that. And in some classes the teacher likes a couple of kids and is nice to them and everybody else can go jump in the lake. That's easy to live with, too. But sometimes a teacher has it in for just one or two people and I never liked that, even before I was one of the one or two people.

Coach McCollough had it in for me, because I wouldn't go out for basketball. He had it in his head that I could be the next McCormick Basketball Hero, another Mace the Ace. He took a lot of credit for Mason's playing, even though Mace didn't really get going good until high school. I would have rather gone out for track, but not going out for anything was the thing that would bug Coach most, so that's what I did. I love to bug people like him.

That day got off to a flying start, me coming in late. Then I had to do more push-ups than anybody else. Then during basketball practice I copied a Harlem Globetrotters routine and got everybody laughing. Then I had to go around the track two more laps than everybody else did. Then I got a lecture on getting a new gym shirt, since mine was torn up pretty bad. That wasn't too surprising, since, like most of my clothes, it'd been worn by Mason before me.

Then Coach got down to what he really wanted to say.

"You know what your trouble is, Mac? You have no competitive spirit."

The way he said it, not having any competitive spirit was like not having the sense God gave a goat. Well, maybe he was right. I don't know that much about it.

"If you didn't have the potential," he went on, "I wouldn't care what kind of lazy turkey you were. But you could be just as good a player as Mason if you'd cut the crap and work at it."

I staggered back in mock amazement, almost knocking Johnny over. "Not that good!" I exclaimed.

Coach clenched his fists, like he was trying to keep from belting me. Everybody held their breath for a second, waiting for him to do it. I thought he'd do that, or tell me to touch my toes for a swat. He didn't swat you like Mrs. Johnson did. Coach's swats would lift you off the ground. For an endless minute we stood like that, then he said, "Showers."

Everybody took off. I did, too, but not as fast as most people.

"Man, Tex, I thought you'd had it," Johnny said later.

I dried my hair off and flipped him with the towel. "Naw. He sounds worse than he is."

"Maybe you shouldn't keep pushing him like that."

I looked up from pulling on my boots. Johnny's freckled face looked serious. "Hey," I said, "what's with you?"

Johnny had been edgy ever since he got grounded, and I had put up enough with edgy people. "You're startin' to sound like Cole."

"Yeah, well what's wrong with that?"

"For Pete's sake, you've been griping about him for days now. Now it sounds like you're starting to take what he says about the evil McCormicks serious."

"Just leave Cole out of this, okay? He's my father, I can bitch about him if I want. Bitch about your own, if you ever see him again."

For a second I really thought I was going to jump up and punch his lights out. It must have showed on my face, because he went charging out of the locker room. I sat there, holding one boot. Having a fight with Johnny was like seeing the sky turn orange. I couldn't believe it had happened.

We didn't speak to each other the rest of the week. When he got his cycle back he didn't come by to pick me up for school. It was serious. I kept up a good front, at school—if he didn't care if we were ever friends again, I didn't—but at home I moped around a lot. I had plenty of people to talk to at school, but just because you know a lot of people doesn't make them your friends. I felt like I did when I found out Negrito was gone. And I had the weirdest feeling that if Johnny hadn't been fighting with Cole, he wouldn't be fighting with me.

I was having lunch at school with a couple of other guys when Jamie came up to the table. "I want to talk to you."

"Sure," I said, ignoring the gibes and snickers from the other guys. If some cute girl walked up to them and said "Frog" they'd have jumped straight up and asked "How high?" on the way.

We moved over to another table that was almost empty.

"I want to know when you and Johnny are going to stop being so stupid."

It was a relief to me that Jamie always said what she thought without hedging around or playing games. But sometimes it took you a little by surprise.

"I don't know," I shrugged, trying to look like I didn't care, either.

"Well, I certainly wouldn't let some dumb

little argument come between me and Linda Murphy."

"I thought Marcie was your best friend," I said. Jamie's eyes were so dark that it always came as a surprise to realize they were blue.

"Oh, that was last month."

"See," I said. "If I went around switching best friends all the time, maybe it'd be different. This time it matters. To me, anyway," I added, tired of lying about it.

"And you think it doesn't matter to Johnny? Listen, he's been acting so weird that Mona has started making him take vitamins. Cole let him get the cycle out early because he's just been sitting around with gloom and doom on his face."

Well, if Johnny didn't want to keep this fight going, and I didn't, you'd think it would be easy to patch things up. But I couldn't see anything easy about it.

"He's going to be out dirt-biking at the gravel pits after school today," Jamie was saying. I wondered when she had started wearing a bra. "And if you two don't quit being so . . . so . . ." she paused, looking for a strong enough word, "asinine, I'm not going to speak to either one of you."

I grinned at her. "Well, that's a real inspiration."

Suddenly she blushed. Turned red clear up to her bangs. I felt my face get hot, and I knew I was turning red, too.

"You know, Tex, you are really cute," Jamie

said. But she didn't say it sarcastic, not one bit. Then she got up and hurried off. I sat there, my face flaming like a bonfire. My heart would stop, then go racing on till I thought I'd suffocate. I had a sudden urge to jump up on my chair and let out a war whoop, but I managed to control myself. For a little while. When Eddie-Joe Cummings came by and cracked a joke, I laughed, and dumped what was left of my chocolate milk on his head.

I got out to the gravel pits with Roger Genet. Roger wasn't real popular with a lot of people, seeing how he was given to stealing things and beating up on kids he knew he could whip. But me and him always got along okay. Anyway, I needed a ride out to the gravel pits, and he had a cycle.

There were five or six cycles out there, roaring up and down the hills, seeing how high they could jump, or who could do the longest wheelie.

Johnny was there but didn't give any indication that he saw me. That bugged me for a second, but then, I knew it wasn't going to be easy.

After a while everybody got together and talked over an old subject, doing an Evel Knievel jump over the creek. Last year a high school senior had tried it, missed, and broke his back. Since then, a lot of people talked about jumping the creek, but nobody really tried it. A couple of people, including Roger Genet, said they had tried it and made it, but unfortunately nobody had been around to witness it. I had always wanted to try it myself, but since I didn't have a bike, I didn't want to take a chance on wrecking Johnny's.

Johnny was saying something about giving it a go, except he was low on gas.

"Hell," said Roger, "that little bitty thing couldn't make it across the creek if it was pumped full."

He had a big old Honda, the kind you couldn't ride legal till you were sixteen. You'd think he wouldn't want to keep reminding everybody he was sixteen and still in the ninth grade, but somehow I don't think Roger ever saw it like that.

"Sure it could," Johnny said. He looked at his fuel gauge. "Maybe I have got enough to try it."

I swung off the back of Roger's cycle. "I don't know," I said, looking at Johnny's fuel gauge. "You look awful low on gas, to me."

I was trying to give Johnny a way to get out of a try, but he looked at me like I was razzing him.

"It's enough," he said curtly, then started up the hill. I took two long strides and hopped on behind him. He didn't say anything. On top of the hill we stopped. The trail led straight down, right to the edge of the creek, then made a sharp left. On the other side the bank was grassy—there weren't any tire tracks there. The creek sides were steep and it was a twenty-foot drop to the creek bed, at least.

"You can get off now," Johnny said.

"Hey, come on," I said. "You don't want to kill yourself."

We looked down to where everybody was grouped, watching. Roger hollered something, we couldn't hear what.

"Off," Johnny said. I got off reluctantly. "Johnny..."

He revved up the engine and took off. I watched him, so antsy I couldn't stand still. Geez, Johnny, faster! I was twisting my fists around like I could change the gears for him. He should have had it wide open by now, full throttle, unless he wanted to be able to stop, unless he thought he'd change his mind... he's going right off the cliff, dammit! I thought. He's going to be dead and I could have stopped him, I should have stopped him... I started running.

Johnny realized he didn't have enough speed, not enough to make the jump, but too much to stop. I felt like I was running in a nightmare; I was going as fast as I could but not covering any ground. Everything was happening in slow motion. Johnny slammed on the brakes and the cycle skidded, turning, but moving right toward the creek. Johnny laid the cycle on its side and they both slid to the edge.

I didn't stop running, even when I saw he wasn't going over.

Johnny was looking at his leg. Most of his jeans and part of his leg was in shreds from the gravel. His jacket had protected his arm, but his knuckles were skinned up, too.

Everybody else buzzed up. Roger had the decency to pause, making sure Johnny wasn't really hurt, before he said, "Run out of gas?"

Johnny didn't look up from picking the rocks out of his leg. I could tell he was wishing he had gone over the bank rather than have to face everybody.

I went over and picked up Johnny's cycle. "Shoot," I said, catching my breath. "He just hit a bump. Anybody could hit a bump. But seeing how he can't give it another chance, I will."

I started up the cycle. I wasn't worried about wrecking it. If that cycle didn't go across that creek bed, for everybody to see, Johnny'd never ride it again anyway. I drove back up the hill, turned, and paused. Everybody was standing to one side, even Johnny had limped out of the way. I saw them for a few seconds, then I didn't see anything but the creek.

When I used to ride in junior rodeos, before money was such a problem, I had the same thing happen to me. You think the crowd is so loud you can't hear yourself think, then you climb in the chute and everything disappears except you and what you're up against. I wouldn't have cared if there were five guys down there, or five hundred, or nobody. I was going across that creek.

I started down the hill, changing gears fast. I didn't even hear the roar of the engine. I kept my eyes on where I wanted to land. A motorcycle needs speed to jump, where it's mostly impulsion with a horse. A horse can tell where you're looking, and head that way, and care if he makes the jump. A horse is a partner, but on a cycle you're all by yourself. Still I leaned and steadied that hunk of machinery like I would a horse coming to a scary jump. When I left the bank and the air whistled around me, and the rocky creek bed floated out behind me, I thought, "Good boy!" and I wasn't talking to myself.

I came down where I'd planned to, but harder than I thought I would. The cycle bounced hard, and we parted company—the cycle going in one direction and me in another. I've had a lot of practice at being thrown from horses, so I know how to relax and roll. And I still got the wind knocked out of me. That sure is a sickening feeling, waiting for air and not really sure you'll get it again.

Somebody came scrambling up the creek bank. I got a mild shock when I looked in that direction. I hadn't cleared the creek by as much as I thought. In fact I'd barely made it.

"Tex?" Johnny crawled over the edge and sat down on his heels beside me. "You okay?"

I nodded, still needing all my air for breathing. Then I tried sitting up. Everything spun around, then settled into place. I waved at everybody watching from the other side. They all cheered and waved back.

"Well, I did it." I felt like I'd won a war, single-handed. "Me and this little bitty thing."

Johnny was getting some color back in his face. He'd been white as a sheet a few minutes before.

"I thought you didn't have any competitive spirit," he said finally.

"I just wanted to see if I could do it," I said.

"It was really great. You looked like a stunt rider or something. I guess . . ." He looked away. "I shouldn't have let Roger psych me out like that. But he thinks that Honda is so cool—"

"Shoot," I said, "I don't see him jumping over here."

Sure enough the others were all driving off in different directions.

"Johnny, there are people who go places and people who stay, and I think we stayers ought to stick together."

He grinned at me. Then he said, "Your jacket's ripped."

I took it off to look at it. It was an old sheepherder jacket of Mason's, but the only coat I had. Ripped was an understatement. It looked like somebody had rubbed a giant piece of sandpaper across the back. Then I looked at the cycle. It lay like a turtle on its back, the wheels still spinning.

"I hope the cycle's okay." I started to get up, then caught my breath. My back was really sore. Johnny got up and give me a hand up. I was tottering around like an old man, holding my hand on my lower back.

"Are *you* okay, that's the question. I can get a new cycle," Johnny said.

"I'm okay."

We got the cycle upright and Johnny tried to get it started, but it'd just splutter and then die. We took turns pushing it home. Johnny limped a lot, and my back ached terrible. We both felt fine.

FIVE

When I got home that evening, Mason was standing against the kitchen doorway, trying to mark off how tall he was. That doorway was full of marks like that. He turned around and measured the distance between the last two marks, scowling.

"Ain't you growing fast enough to suit you?" I asked. I tried to take my jacket off without him seeing it—I didn't need a lecture about how I wasn't going to get a new one. Luckily he was too busy writing on the wall to notice me.

"Shoot," I said, "you're over six feet tall now. You don't want to turn into a freak."

"I'd turn into the bearded lady if it'd get me into college," he muttered. He turned to me. "You think I've stopped growing?"

I looked at him for a minute. You can tell

how tall a horse is going to be by the length of his legs when he's young; I figured that'd work for humans, too. "Naw. You got a few inches to go."

Mason sighed with relief.

"Okay, I told you something, now you tell me something. How do I look?"

Mason choked on a laugh, then really looked at me for the first time that afternoon. "Well, you look pretty messed up right now. What happened?"

"I jumped the creek on Johnny's cycle and didn't land right. I mean, do you think I'm good-looking?"

"The creek? You jumped the creek? The same one Joe Taylor smashed himself up on?"

"Yeah, but I made it okay."

"You hare-brained idiot, you're not going to make it to sixteen if you keep doing dumb—"

"Okayokay. But Mace, just imagine, man, how it's going to freak everybody out at school, that I jumped the creek and made it. They are going to go nuts."

I was thinking about what Jamie would say. She already thought I was cute—dumb word, but girls used it . . .

"I am the one who is going to go nuts. Goddammit, Texas, you could have ended up in the hospital for the rest of your life!"

He broke off suddenly, like he was too mad to even talk. He stalked off into the kitchen. In a little bit I followed him.

"Hey, look, I won't do it again, okay?" I said.

"No, but you'll do something else just as stupid. Sometimes I don't even know why I try."

A flinch ran across his face. I slunk off into the bathroom to wash up. I really didn't mean to get Mason upset like that. Pop wouldn't be upset when he heard. He'd probably think it was really cool.

I wished I could ask Mason if he thought I was good-looking. It's hard to tell about your own face, since you see it all the time. I do have really good hair. It's probably my best point. It's a light bright gold-brown, the same color as my eyes. It sun-bleaches even lighter in the summer.

I stared at the mirror, trying to see what somebody else would see if they didn't live in that face. A chipped tooth. That scar over my left eyebrow that Mason had given me. I have kind of a boney face, like Mason, but where he looks like a proud hawk, I look more like a surprised antelope. I've taken too much razzing about my dimples not to know I have 'em, but I can't tell if they're a good thing to have or not.

"Nobody's going to mistake you for Robert Redford, if that's what you're worried about."

Mason's voice made me jump. He was sitting on the bed changing into his jogging tennis shoes. I felt myself get red.

"I ain't worried about it."

"You ought to get your hair cut." Mason had cut his short last summer, and now a lot of kids in his class started cutting their hair shorter, too. If Mason wore a pickle behind his ear, there's people who'd do that, too. I can tell he gets a kick out of that, even though he never says anything about it.

"Jamie likes my hair just like it is."

Mason raised his eyebrows, but didn't say anything. He looked like he was pretty much over being mad at me.

"I'm going into the city Saturday," he said. "You need anything?"

"Hey, I want to go, too," I said. "Come on."

"Okay, okay. But you'll have to get up early, I got a nine o'clock appointment."

"Sure," I said, thinking about how I could talk him into letting me drive some. Mason wasn't too bad about letting me drive, not bad at all considering I didn't even have a beginner's permit, but right after jumping the creek wasn't the right time to bring it up. I'd let him drive in and pester him to let me drive back. He'd probably be ready to let me; city traffic always made him nervous. It didn't bother me none.

"Maybe we can go to a movie," I said, getting excited. I hadn't been to the city for months, except for going to the Fair.

"Maybe," Mason said.

"And McDonald's."

"Sure," Mason said. "Sure."

Saturday morning I got up even earlier than Mason. It was pitch black outside. He was awake, but he didn't seem to be in any big hurry to get up, which was funny after the way he'd pushed me to bed the night before, carrying on about how he couldn't be late. I wasn't sleepy like I was on school days. I fixed some coffee and scrambled eggs and burned three pieces of toast before I finally gave up on it. I heard Mason splashing around in the tub, but I'd taken a bath the day before. I wasn't about to take another one.

"Hey, let's go to Westmall," I said, as Mason came into the kitchen. Westmall was a neat shopping center, all enclosed like a giant cave. I loved walking up and down looking in all the stores. Mason hated it. He always said the place made him feel like a hick because we couldn't afford to buy one single thing there. That wasn't true—there was a big discount store at one end of the mall where we'd bought jeans and shirts and stuff lots of times. Mason could get the weirdest ideas in his head.

Sure enough he said, "I don't want to."

When I opened my mouth to argue, he went on, "You can go by yourself while I'm at the hospital."

"What you goin' to the hospital for? Visiting Charlie Collins?"

I figured Charlie might be recommending him to some college or something. Charlie knew a lot of people.

"I'm going in for some tests."

For a second I thought he meant tests like school tests. Then I realized he meant tests like a doctor gives you to see what's wrong.

I set my coffee down so hard it splashed out and burnt my hand. Mason got up and started washing the dishes. He hadn't eaten a bite of anything. I just sat and stared at him, my insides quivering. Finally he looked over at me, "Good God! What's the matter?"

"Is it cancer or something?" I asked shakily. We didn't watch much TV, but it seemed like every day at school all the girls would be talking about some show where somebody, some kid or football player or hippie person, was dying of cancer. "It was *so* sad," they'd say. "I just *cried*."

Anyway, I knew sometimes people went into hospitals and never came out. What if Mason was dying!

"Tex," he laughed, "it's not cancer. At least I'm pretty sure it's not. I just thought I'd better check it out."

I just looked at him. Maybe he was putting up a front. He never could stand anybody feeling sorry for him. "Yeah?" I said. "How sure?"

"Well, Charlie didn't think it was serious. He set up my appointments for me."

And when I didn't look convinced he said, "I don't lie to you, remember?"

I licked the back of my hand where the coffee had made a big red blotch. Maybe it wasn't serious. He looked healthy enough, a little on the lean side, but I was, too, and there sure wasn't anything wrong with me. "If you ain't sick why ain't you eatin' anything? It ain't bad cooking, for me."

Mason shook his head. "I'm not supposed to eat anything until the tests are done. Come on, get your teeth brushed, it's time to go."

It looked like everybody in the county decided they had to be in the city by nine o'clock. Pretty soon Mason was gripping the steering wheel so hard his knuckles turned white. Then he was either muttering to himself or hollering out the window. Being a jock he knew some good things to holler.

"I don't know why the hell people don't obey the law!" he snapped, as a big silver Chevy roared past us. I didn't point out the fact that we were doing ten miles over the speed limit ourselves. I just didn't have the heart to laugh at Mace like I usually did. Maybe he really was sick. After all, people didn't go to the hospital to get tested for measles. If Pop knew Mason was sick, he'd come home. He'd do anything for Mason. But if Mace had to stay in the hospital, how could I get hold of Pop to let him know? What was I going to do if something happened to Mason?

I stared out at the oncoming cars.

"Mace," I said, struck by a thought, "did you ever think that all those people in those cars have a whole separate story to them, that it's just as important to them as our stuff is to us, and we don't know anything about it. Maybe sometime we'll run across somebody and two years ago they were driving past us on the highway and we never knew it. Like sometimes we meet people and bump off of them and never see them again and we never know why paths cross."

"Yeah, sure I spend a lot of time worrying about that," Mason said sarcastically.

"I'm not worrying, I'm thinking."

"Well, don't pass out from the shock. *I'm* thinking it's damned dumb to have a two-lane highway in this day and age." He leaned out the window and yelled some more.

The traffic was a little better when we got to the expressway. It's better on Saturdays. The hospital was almost clear downtown. Mason was in a real sweat by the time we got there. He pulled down a side street and parked.

"Why don't you go on to the shopping center and meet me back here in a couple of hours?" he asked.

"Don't you want me to go with you?"

"No," he said bluntly.

"How come?" My voice was rising. "You don't want me to know you got something serious, right? That's it, ain't it?"

Mace reached over and patted my shoulder. "Calm down. It's just that I'll have to be doing a lot of waiting around and it'd get on my nerves to have you fidgeting around with me. You don't sit still easy, you know."

He got out, and I slid over to the driver's seat.

"Mace . . ."

"Come on, kid, I'll be okay. Look, be careful, and don't drag race."

"Okay."

"And don't lose the car."

Once we had lost the car at Westmall because the parking lots were so big. We'd thought for an hour it'd been stolen, when it was just parked on the other side of the mall. I'd have to make sure of which door I went in.

"Okay," I said. "You want me to get you anything while I'm there?"

"Oh, sure," Mace said, "here's my American Express card, get anything that catches your fancy." Then he turned around and walked toward the hospital. I watched him for a minute, then started up the truck and drove off.

I got to the mall without any trouble, except I sort of ran a stoplight, but I didn't get caught. Most of the stores weren't open yet, so I walked up and down looking in the windows. After an hour or so, it started filling up with people, but I still felt kind of lonesome. It was weird being lonesome in a place full of people. I never get lonesome out hunting or fishing by myself. I tried not to worry about Mason.

I went into the bookstore to see if they had a copy of *Smoky the Cowhorse* in paperback, but they didn't. I spent a while looking at the books anyway. I like to read okay, but I just never seemed to have time for it.

Then I went into a jeans shop. I can't believe people buy jeans already faded, when just a month out riding in them will do a plenty good job.

"If you can help me I'll let you know," a voice said.

"Okay," I said absently, then realized that wasn't how the sentence was supposed to go. I looked at the salesgirl.

"Sorry," she smiled, "I just get tired of saying the same thing all the time. Anyway, I wanted to see if you were paying attention."

I was now. She was really cute, with gold hair and big light-green eyes and the neatest gold freckles all over her face. She was maybe a year older than me.

"Would you like to try on some jeans?"

"Uh, no, I'm just looking," I said. Somehow the thought of picking out jeans with a girl salesclerk was kind of embarrassing. Then I felt dumb. I must look really dumb. Maybe this was why Mason hated going into stores, they made you feel dumb. I turned around and nearly knocked over a rack of shirts. I set them up and hurried out.

I heard her say to another salesgirl, "I knew that kid wasn't going to buy anything."

"Yeah, but he was a foxy little devil, wasn't he?" the other one said.

I was so confused I couldn't see straight and almost knocked down an older lady carrying a bunch of sacks. I hate getting embarrassed. The more you think about it the worse it gets.

I stopped at the snack bar to get a Coke to cool off. That salesgirl was kind of cute, too. I couldn't figure out why girls were getting so cute lately. Last year they hadn't been so great.

I sat down on the edge of the indoor fountain and tried to calm down a little. I couldn't go charging around knocking things and people over. That would really look dumb.

Every once in a while somebody would come along and toss a penny in the fountain. The bottom was covered with pennies. That always made Mason so mad he couldn't see straight. Shoot, if it'd get me a wish, I'd try it myself. Anyway, it wasn't my pennies they were throwing. Try and tell Mason that.

Then I felt mean, thinking about Mace like that, and him at the hospital that very minute.

I was tired of walking up and down, so I went outside. Across a stretch of parking lot was a big sporting goods store, a movie house, a record shop, and an ice cream parlor. If you didn't know better, you'd think stores multiplied like rabbits.

I walked over to the sporting goods store. I meant to look at the guns, but got sidetracked by the fishing stuff. It had been a long time since Mason and me had gone fishing. Maybe if I got us a new lure, he'd want to go again. I like hunting better, myself, but fishing was the only thing Mason really relaxed at.

I looked the lures over, trying to remember what ones he already had, and trying to find one that didn't cost too much. I couldn't believe the way they had gone up in price. Finally I decided on one; it would take what was left of my money, but I hadn't planned on buying anything anyway. Mason could pay for lunch.

Then I went to look at the shotguns. There was no way I could get one, but I liked looking at them. I found a 20-gauge I really liked. I put the fishing lure in my shirt pocket and picked up the gun to see how it balanced. It seemed a little stock-heavy to me, but maybe that was just sour grapes since I couldn't get it. Good sights on it, though. I sighed and put it down. Duck season was coming up . . . well, Christmas was coming up, too.

I walked around a little more, looking at tennis stuff and skiing stuff and wondering if there were enough people to buy all the stuff in stores. I guess there are, though, or there wouldn't be so many stores. I looked at the water skis. I went water skiing once, and man, I loved it.

Someone tapped me on the shoulder. I turned around and faced a salesman.

"Kid," he said, "can you read that sign?"

I looked where he was pointing. "Shoplifters will be prosecuted."

"Sure I can read it."

"Do you know what it means?"

"Yeah. You catch somebody stealing you'll do your best to send 'em to jail."

"Good. Now come with me."

I did, not even thinking about asking why. When you're in a strange place you don't think about having control of anything.

I followed him into a back office.

"I got another one, Ed," he said to a man behind a desk. Ed looked up at me wearily.

"What was it?"

The salesman reached over and pulled the

fishing lure out of my pocket. For a second I didn't know what was going on. Then I broke out in a cold, sick sweat. They thought I was stealing it!

"I was going to pay for it," I said, when I could get my breath. I went from shocked to mad to scared, so quick I couldn't tell which I was feeling. "I was going to pay for it," I repeated, trying to keep my voice level. I sounded guilty. I even felt guilty. I must look guilty, too, I thought frantically. All kinds of visions were going through my mind—what would Mason say, what would Pop think, Jamie . . . being put in jail. I couldn't go back to jail!

"They all say that," Ed nodded.

Say what? I wondered, then I remembered saying, "I was going to pay for it."

"I never stole anything in my life!" I said. Borrowing that car hadn't really been stealing. Anyway, I was just twelve when that happened—

"Now that one I never heard before," Ed said.

I was so scared and sick I was close to crying. They'd never believe me. Nobody'd ever believe me. Pop would disown me. Pop was funny and talkative and loved to laugh and tell stories, but he had never ever said a word about prison or what it was like, just drilled it into us to respect the law. That probably impressed us more than if he had given us a day by day description of it. Anyway, I knew what it was like, sort of. I remembered sitting in that jail cell, listening to somebody beat on the bars down the hall, screaming "I'm drunk and I'm proud of it!" over and over again and the smell was so bad and the walls got closer and closer and I knew if I had to stay in there I'd go nuts like a caged animal, and beat my head in against the wall, and I was trying to sit real still to keep from doing that when Pop came to get me. From the look on his face I thought I'd go straight from jail to an orphanage. And when we got outside he'd belted me; the only time he'd ever hit either of us.

"I just put it in my pocket for a minute while I looked at a gun," I said, as steady as possible. The way they were looking at me made me feel like I was the worst kind of trash. Mason, even if he believed me, he'd never live this down. He had such a pride thing about never getting into trouble when he had so many chances to. He had turned very sarcastic on Johnny once, when Johnny smuggled a carton of Eskimo Pies out of the Safeway store, made him so miserable he couldn't even eat more than one or two.

"Was he out of the store?" Ed asked the salesman. "Or about to leave the store?"

It was weird, being talked about like you weren't even human.

"He was inside," the salesman said reluctantly.

"Turn him loose. This time." He went back to his papers.

The salesman opened his mouth to protest, then snapped at me, "You heard him, get out."

Everything had happened so fast I still didn't feel like I could move. "I was going to pay for it." I made one more effort to clear myself. "I got the money to pay for it." I almost took my billfold out to prove it.

"Kid," Ed said, "people come in here, kids with bigger allowances than my salary, and it's just like mountain climbing to them. They take things because they're there. Sometimes we even find merchandise in the trash bins—once they get away with it they're bored with it."

He glanced up at me, and for a second I felt like he believed me. "Still want the lure?"

I looked at it, laying there on his desk. A fishing lure was going to make me slightly sick for a long time to come. I shook my head. He escorted me out, but he didn't need to worry. I was never going into that store again.

I stood outside for a few seconds, still a little dazed.

It took me a few seconds to realize somebody was calling my name.

"Tex—are you deaf or stoned or what?"

I turned around. Jamie was standing outside the ice cream parlor with three or four other girls.

"Hey," I said, "hi."

She was wearing jeans and a shirt I knew belonged to Johnny and that he'd told her not to touch. It was a little too big for her. I couldn't tell if she was wearing a bra or not.

"What are you doing here?" she asked. The other girls were giggling and whispering together.

"Uh, nothing. Just killing time while Mason's at the hospital." Just looking at her was making me a little dizzy. The jeans salesgirl wasn't anything compared to Jamie. I wondered how I could have ever thought any other girl was cute. I must be in love, I thought, and my insides got shaky and my hands started sweating and there was a buzzing in my ears that made it hard to hear.

". . . doing at the hospital?"

I realized Jamie had asked a question. Hospital. Mason. Oh, yeah, Mace was at the hospital. Huh. Maybe he wouldn't want everybody to know he was sick or something. Mason was really a fanatic about keeping things private.

"He went to see your brother Charlie. So Charlie could show him the hospital. Mason's thinking about being a doctor."

Then it occurred to me that she might talk to Charlie and find out different. Somehow when I was around Jamie I didn't exactly know what I was doing.

"Going to make money and nurses, huh?" Jamie said. I couldn't think of anything to say to that.

"Well," she said, "that's what Charlie says about being a doctor. Cole'd kill him if he heard. Of course, Charlie's only kidding. Partly."

Jamie's eyes were bluer than a twilight sky and her hair had a sheen like a summer horse, and I wanted to squeeze her till she crunched—I wanted to rescue her from something . . .

". . . spent the night with Laura," she was pointing to one of the girls, I couldn't tell which one, they all looked alike.

"We're going to the movies, and Cole's going to pick me up later."

I looked up at the marquee. "I hear that's a good movie."

I felt like everything I was saying was coming out really dumb, but I couldn't seem to stop it.

"I've seen it before. I like the guy who plays the smuggler. He is a doll."

Anger flashed over me like a freak lightning storm. "Oh, yeah? I hear he's a real wimp."

When the covey of girls giggled, I realized how that sounded. It sounded jealous. I might as well have yelled "I love you" over a loudspeaker. That was how it sounded. I felt the red spreading up my neck till my ears got hot. I expected Jamie to be glaring at me, but she just said, "Whatever you say." Then I noticed she was a little pink, too.

All I could think of was getting out of there before I said anything dumb again. "I got to go," I said. I did, too, it wasn't just any excuse, Mason would be waiting.

"Well, I'll see you around," she said.

"Yeah," I said and hurried off. My heart was thumping in my ears so hard I thought I would be deaf, but I heard, clear as anything, one of those girls say, "Oh, Jamie, he's cute," and Jamie say, "Yeah, he's real sweet, too."

She liked me! I had a chance. I stepped out

in front of a green Firebird and nearly got run over. It gave an obscene blast of its horn and an angry squeal of its tires, but I didn't care.

I had a chance.

Mason was waiting for me at the hospital. He didn't even offer to take the wheel. I was still so wrapped up in Jamie that I'd driven three or four blocks before I remembered he had his problems, too.

"Did you find out anything?" I asked finally. He sat slumped in the corner, glowering like a cornered wildcat. I half expected him to shoot off sparks.

"Yeah, I found out something. I got a goddamn ulcer, that's what I found out. I got a goddamn ulcer. Anything else you want to know?"

That wrenched my mind off Jamie. "An ulcer? Is that bad? I mean, serious?"

"Well, it could get serious. It could get pretty goddamn serious." His voice was shaking. I kept quiet, and drove as careful as I could, and by the time we got to McDonald's he'd straightened up, took a couple of deep breaths, and unclenched his fists.

I ordered two Big Macs, two orders of fries, and a chocolate shake. Mason had a strawberry shake. That was all.

"Have some french fries," I said, holding them out.

He shook his head. "I'm not supposed to eat

stuff that'll bother me. Last time I had french fries they bothered me." He was quiet a second. "There goes the ol' chili."

I stared at him. No more chili? He'd starve! "Hey man," I said, "I'm sorry."

"You and me both. I got to get some pills, too . . . I'm not supposed to let things get to me." His voice broke. It was awful to see Mason upset. He always stayed cool. "Now just how the hell do I do that?"

I figured I was one of those things that got to him. Me and—"You can still play basketball?" I asked anxiously.

"Yeah, yeah." He nodded. "So far. It could be worse."

"Have a hamburger. You can have a hamburger, can't you?"

"Yeah, but I'm not hungry. I had to drink some barium and it really killed my appetite."

Mine was gone, too, but I finished my second Big Mac and cleaned up the fries anyway. I don't get to McDonald's very often.

"It could be worse," he repeated a couple of times, and by the time we were ready to go he sort of seemed to believe it.

"I hope these damn pain pills work."

"You been having pains?" I said, surprised. I never noticed it.

"I didn't publish it in the paper, but yeah, I had 'em. Bad, too. I was puking blood after practice last week."

I looked at his lean, hawk face. There was

something about him hurting like that and being too proud to say anything that really made me sad.

"You want to go see Lem?" he asked, like he was changing the subject.

"Sure," I said. "I'll drive."

"Okay," he said, once we were driving down Peoria. "What happened to you today?"

"Nothing much. I looked around the stores."

I decided that hearing I'd been accused of shoplifting wouldn't do his stomach any good. So I skipped that. I didn't need to tell him about knocking over a rack of shirts, he'd laugh. So I skipped that. I had some other stuff to talk to him about, so I skipped the part where I saw Jamie.

"Nothing, huh? Well, that's a record, for you."

I felt like I'd told the biggest lie. I'd gone through ten times as many emotions in that one day than I usually did in a week at home.

I said, "Mace, when you're making out with a girl and everything, how can you tell if it's okay to go further?"

Mason looked at me. "I didn't know you'd even got to the kissing stage yet."

"Well, I haven't. But just in case—"

"You know, that's all. If you make a mistake, she'll let you know." He paused. "I hope you're not going out for the school stud."

"I was just wondering, that's all. Is it really neat, going all the way?"

"Kid, don't get started on that now. Once you get going with girls you can't quit. Anyway," he added quietly, "I can't tell you what it's like. I never done it."

I almost drove into the wrong lane.

"Don't tell anybody," he warned in a voice that promised instant death.

"But you and Bobby, you said—"

"Hell, Bob hasn't made it any further than I have. But for different reasons. Everybody talks. And if you don't know enough not to believe everything you hear, especially in a locker room, I'm telling you now. I never could stand the thought of getting tied down," he went on. "I never wanted anybody to have any kind of a hold on me. Look what happened to Lem. Nobody is ever going to stop me from getting out of here."

Well, I knew that. But, boy, that was desperate! It was almost scary.

"It hasn't been easy. And don't think I haven't had plenty of chances."

"Sure," I said. Lord, I knew he had chances. Being the school hero gave a guy chances.

"When I get to college, and at least have that much . . . if I can get over the feeling I won't be trapped . . ."

His voice trailed off. We were quiet for a long time. I was so surprised about all this I couldn't even think of what to say. A whole lot of girls liked Mason. He wasn't what anybody would call real handsome, but there was something about him that girls really dug. And he'd been dating since his sophomore year. But probably he'd never been in love. I figured that'd make a big difference.

"If you're thinking about Jamie Collins you might as well forget it," he said suddenly, and hateful, like he was mad he'd told me so much.

"What makes you think it's Jamie?"

"You said her name in your sleep the other night and it was obvious you weren't having a nightmare."

I went red, then mad. "Well, why not Jamie? It's 'cause the Collins got money and we don't, right? You think money means everything."

"Nope, kid," he said gently. "It's because you're Tex and she's Jamie. Money has nothing to do with it."

What the hell did he know? Where did he get off telling me this stuff? Him and his damn honesty. "Why do you think you got to be so damn truthful? Why can't you just tell me a nice lie once in a while?"

"Texas, all my life I wanted somebody who knew more than I did to tell me the truth. I really wanted that. I never got it. I had to learn it all the hard way. I'm just giving you a present I always wanted."

"Thanks a lot," I said.

SIX

We stopped by a discount drug store and picked up Mason's prescription. I looked at magazines while Mason called Lem to get detailed directions to his place. It was easy for us to get lost in the city. I couldn't make up my mind to look at *Playboy* or *Western Horseman*, so I looked at both. While I was reading about studs and halters and hindquarters in one magazine, pictures from the other kept flashing through my mind. I about choked, laughing. I know the guy behind the counter thought I was on something.

Lem lived in an apartment in a part of town that was blocks of apartments. I don't see how people keep from getting lost in it. The apartments were called Southern Ivy and he lived in

Southern Ivy II. That was the section for people with kids. We walked through rows and rows of apartments.

"This is it," Mason said. I'd take his word for it, but how he could tell this doorway from any of the others, I didn't know. I tried looking in the window, while Mason pushed the door buzzer, but the curtains were pulled shut. There was some scurrying around inside, like mice in a barn, then Lem's voice said, "It's all right, it's them," and the door was unbolted.

"Well, hi!" Connie said. She was wearing hip-hugger jeans and a short pink sweater. I thought girls were supposed to get fat after having a baby, but Connie looked the same as she used to, maybe even a little curvier.

"Come on in," Lem said, slapping each of us on the back.

It was a really neat apartment. I mean, nice. It wasn't exactly neat, because clothes and diapers and stuff were laying all over everything. But I was used to messy housekeeping at home.

"Have a seat." Lem shoved a bunch of clothes off the couch.

"Want to see the baby?" Connie asked. She ran upstairs to get him before we could get "yeah" out of our mouths.

"How about a beer?" Lem asked. Mason said, "No," and added, "he doesn't either." He was just trying to be a big shot, as usual. He knew I don't even care for beer anyway.

"Coke?"

"Sure." I followed Lem into the kitchen. They had a dishwasher and refrigerator and everything. Lem gave up looking for a clean glass and handed me a cold can of Diet-Rite.

"How you been, kid?"

"Fine," I answered. Something was different about Lem. I couldn't quite tell what it was. He looked out of place. Taller and heavier than Mason, rangy as a steer, he looked just plain clumsy in this little kitchen. I thought about the time we were training his Appaloosa colt, a year back. He hadn't been clumsy then.

When we went back to the living room, Connie was trying to make Mason hold the baby.

"Well, here, Tex, you hold him. He won't bite you." She handed him to me. I took him. It wasn't all that different from holding a puppy or a kitten.

"Shoot, he's not much heavier than a basketball, and I never seen you drop one of them," Lem laughed at Mason.

The baby was blonde like Connie, but had dark eyes like Lem. And it was funny to look at him, because Lem had real heavy, bushy eyebrows, like furry caterpillars, and that baby's eyebrows were way too dark for a baby. It looked like somebody had pasted them on him.

I wondered when I had a baby if it'd look like me or Jamie.

"Wow, I bet you have fun training him," I said.

"A kid is not a colt," Lem said.

"In most cases," Mason said, looking at me. I

gave the baby back to Connie. I didn't mind holding him, but it occurred to me that he probably wasn't housebroken yet.

The phone rang. Lem went to answer it, while Connie sat on the floor and talked silly to the baby, calling him "mother's precious darlin' " and junk like that. I hoped the baby wouldn't grow up thinking he had an idiot mother.

"Yeah, I got it."

We could hear Lem, even though we weren't trying to. "The Holiday Inn? Sure. What time?"

"You guys see my new car?" Lem asked, after he hung up.

"You got a new car?" I asked.

Mason said, "Gas jockeys must be getting better wages these days."

Lem gave him a funny look.

"Honey, you take them for a ride," Connie said cheerfully. "I'll stay here in case the phone rings."

The car was a dark blue Pontiac, with white insides.

"Wow!" I said. "It's got air conditioning! Push-button windows? Come on, let's see how fast it'll go."

We drove out to the expressway so Lem could get it going fast. I had fun running the windows up and down. Mason was quiet.

"I had an eight-track tape deck in here," Lem said. "But it got ripped off the first day. It's a bad neighborhood for things getting ripped off.

But not bad for getting mugged or anything like that. Those apartments are brand new. Only a couple of chicks lived there before we did."

Mason was still quiet, even when Lem took it up to ninety-five. He would have never sat still for me doing that. Then he said, "What you dealin' in, Lem? Grass, speed, horse?"

I was so shocked I couldn't believe I was hearing him right. But Lem seemed to know what he was talking about, because after a minute he sighed and said, "Mostly grass, and a little speed."

Mason swore. Lem got a defensive look on his face. "Don't get high and mighty with me, Mason McCormick. It's a lot of easy money. I'm just a go-between, it ain't like I'm selling it out on the streets or somethin'. I mean, I know who I'm dealin' with."

"Sure." Nobody could sound as contemptuous as Mason. It made me cringe. "I reckon you don't use it, either."

"No, I don't. Well, maybe a couple hits of speed before a concert or a movie or somethin' like that, just to make it better, get your money's worth, but I ain't hooked on it."

"How about Connie?"

"She's real careful, you know, so one of us can stay straight for the baby. She uses a little speed to keep her weight down, but shoot, Mace, the doctor would give her that. Anyway, I've seen you as high as anybody else—what about that party at Joe Ray's place?"

I didn't dare open my mouth, even though

Mason's always tried to give me the impression he's as pure as the driven snow when it comes to drugs. I could see he was in a mood to slug somebody.

"I never set up shopkeeping in it. What about the baby? When he's older? He's gonna think they're vitamins."

"Shoot, no. When he gets old enough to know what's goin' on, we'll quit."

"You are really stupid," Mason said.

Lem's jaw line went hard. "Yeah? Well, it's a little easier for you. You don't have a wife and kid lookin' for you to take care of them. I want a nice place to live and food for the baby and Connie's so pretty, it's hard for her to do without clothes and stuff. Who am I hurtin', anyway? It'd be going on whether I was in on it or not. Somebody else would just be making the couple of hundred bucks a day. And that, buddy-boy, is what I'm making. I guess you could turn that down, couldn't you?"

"Yeah, I could," Mason said, but I could tell he was just sick at the thought of all that money.

"Well, maybe you could. You don't have people depending on you."

"What do you think he is?" Mace jerked a thumb toward me. "A baboon?"

"It ain't the same. Next year you'll be gone and Tex'll have to learn to take care of himself. A kid is for life."

We took an exit, went under the expressway, got back on going in the other direction.

"Shoot, Mace," Lem said finally, "there's no need to get into a big hassle about it. Everybody is doin' stuff like this, I'm just into a little more than most people. I could tell you about some guys in Garyville . . . it ain't any big deal."

"You'll think it's a big deal when you end up in McAllister," Mason said shortly, and Lem didn't try to talk to him again.

When we got back to the apartment, we stood around by the car, awkwardly, not knowing what to say.

"You guys want to go get a pizza or something?" Lem asked finally.

"No," Mason said, "we better be gettin' back . . ."

I couldn't think of anything we had to be getting back for, and I could tell Mason couldn't either.

"Well, keep in touch," Lem said, and stuck out his hand. Mason shook it, but didn't say anything. I had a feeling he didn't want to keep in touch. In fact this might be the last time he saw Lem.

"See you kid," Lem nodded at me.

"Yeah," I said. I had a knot in my throat. Lem wasn't the kind of person you could cry over—but still he had always been around, always been part of my life, and it's hard to let go of a part of your life.

"You can drive," Mason said. He figured that would take my mind off Lem, and it did. A little bit.

* * *

"I always knew Lem wasn't the brightest guy I'd ever known, but I never figured him for an absolute moron."

Mason wouldn't change the subject, no matter what I'd say or do, even fifty miles an hour in a thirty-five zone.

"See what I mean?" he said suddenly. "That's what messin' around with a girl'll get you. He's never going to get anywhere except prison, the way he's going now. He is just good and stuck, even if he did get out of Garyville. Man, I am never going to get stuck anywhere!"

I didn't think he was right, because if you were where you wanted to be—even married and a daddy and in Garyville—you weren't stuck, but I never was as good at arguing as Mason. But he did have kind of a point. I mean, the year Lem and Connie were going together it was the romance of the century, even kind of Romeo and Juliet, since their folks did everything they could to break them up. There was always some big dramatic scene going on, with Lem threatening murder and Connie running away. Everybody thought it was the neatest love thing going on in town. But nobody thought about them anymore, because that was last year. And Mason was right about one thing, Lem didn't seem too happy.

"God, is he dumb!" Mason said for the hundredth time. I saw a guy thumbing a ride up ahead, and figured it would be a good way to finally switch the subject. I swerved over to the side of the road. He was a young guy, not a whole lot older than Mason, small built, wearing sunglasses and clothes that didn't seem to fit him too well. There was something funny about his hair. It was a muddy brown color. But the color didn't stop at his neck, where his hair did, it splotched on down into his collar. He must have dyed his hair, I thought. Weird.

Mason slid over to the middle to make room for him.

"Where you guys headed?" he asked, slamming the door.

"Garyville," I said.

"Well, that's the right direction."

"How about you?" Mason asked.

"State line."

He was from around here, it's usually pretty easy to tell an out-of-stater. He kind of reminded me of somebody, but I couldn't think who.

"I wish you guys were going further than that," he said. Now there was another weird thing, the way he said that. Like a statement, or an order, not like a request. Suddenly I felt Mason freeze. He didn't move a muscle, or say anything, or change the expression on his face. He just quit breathing.

"I think it'd be nice if you kept on going to the state line," the hitchhiker repeated, and when I glanced over at him, I saw he had a gun pointed at Mason's ribs. He kept on staring ahead of us, down the highway, a funny kind of little grin on his face, and I saw that I'd been wrong about him being not too much older than Mason. You don't have to live a long time to be old.

"Just take it easy," he went on. I took it easy. There didn't seem to be anything else to do. I wasn't scared. I didn't want him to shoot either one of us, but so far he hadn't, and I tried keeping my mind on the road.

"Isn't this a little extreme?" Mason said in his coolest voice. "I mean, sooner or later *somebody* would have been goin' to the state line."

The hitchhiker laughed. It was amazing how normal he sounded, like Mason had just told him a funny joke. "Yeah, well, I couldn't wait till later. I got people looking for me. I could have been long gone, by now, but I thought I had business. Yeah," he repeated, "I had business to take care of."

After a minute of quiet he said, "It's loaded, in case you're wondering."

"There was no doubt in my mind," I said firmly. I was trying to think of something to do—run the truck into the ditch, speeding so a cop'd stop us—but nothing seemed safe enough.

"I'll be up for murder one, as it is now," the guy said. I didn't want him telling us too much. I figured the less we knew, the better, but he was in a talkative mood, and I wasn't in a position to say "shut up."

"So I haven't got a thing to lose. Keep that in mind. Not that I like killing people. Both times it was a case of have to. I'd like to keep it that way. I'm beginning to see how some get to liking it, though. I can see that."

I could see Mason's face in the rearview mirror. I was glad he wasn't driving. His freckles were standing out like 3-D.

"You guys ever been in prison?"

"Nope," I said.

"Well, I'm going to give you a piece of advice. Kill somebody as soon as you get there. Course if you're a big guy, or old enough, just beat the crap outta somebody. But if you're young . . . kill somebody right off the bat. Then let 'em know you'd do it again. People'll leave you alone, then. That is the way to make high society, in the joint."

I was thinking about Pop being in prison. No wonder he never talked about it. Locked up with people like this—how did he ever stand it?

"I guess this isn't doing much for your ulcer, huh?" I said to Mason. He managed something that passed for a grin, and shook his head.

"Shut up and drive, cowboy," the hitchhiker said. He didn't want us to make a move he didn't okay.

We drove on through Garyville; the highway runs right through the outside of town. We drove past a dozen people who knew us; kids hanging out in front of the car wash, sitting on their cars or in them; a bunch of people honked or waved at us.

"Somebody gonna think it's weird, a stranger in the car?" the guy asked.

"No," Mason said, and cleared his throat, "we pick up hitchhikers all the time."

"I think we're going to give it up, though," I said, and the guy laughed his head off.

"I should have been gone a day ago." He started up his monologue again. If one person does all the talking, that's a monologue. That's one thing Miss Carlson taught me. If two people are talking it's a dialogue . . . I wondered what you'd call one person talking and two persons sweating. I wondered . . .

"But I thought I had business. Waited for it for years, the big revenge trip, and when it came right down to it, it was nothing." He was talking to himself, not us, like we didn't count. Like we were dead already. ". . . him lying there looking up at me, and he says, 'Get it over with,' and it was like all the air out of a balloon. All these years of planning, waiting to dig the look on his face, and then I just didn't feel like finishing it. Not that I gave a damn about the son of a bitch—I was just too plain bored. And to think I could have been planning something constructive all this time, like the quickest way out of the country . . ."

A highway patrol car had been behind us for the last mile. I didn't know how to get Mason to glance in the rearview mirror without tipping off the hitchhiker. I increased the speed just a little, and the guy said, "A mile an hour more and he's dead."

Mason grunted as the gun was jammed into his ribs. I slowed down.

"And I know about Smokey, too, so just stay cool."

I swallowed a sigh and waited for the patrol car to pass us. I've never been driving the speed limit yet that a highway patrol car didn't scoot on by. But it just stayed behind us, a few car lengths behind us, probably just waiting for the chance to pass. We were on a two-lane highway, going around bends and up hills. There wasn't a whole lot of chances to pass.

I was getting fed up. I hate being nervous, it ain't a natural condition for me. If it'd been just me in the truck with the creep, I wouldn't have been so tense. But Mason was giving off tension like sound waves, and I could just imagine what was happening to his stomach. It really bothered me to know that he was scared.

This might be my last day alive. It hadn't been exactly the most fun-filled so far.

I said, "I think he put his lights on."

The hitchhiker turned to look. I put my foot on the brake and tried shoving it through the floor. I spun the steering wheel like it belonged to a boat instead of a car. The truck whirled around and slid like a panicked horse. It skidded across the road and teetered on the edge of the ditch for what seemed like an hour; every single thing that had ever happened to me flashed across that windshield like a movie. It almost turned over, then rocked to a slanted standstill, half in the ditch. The hitchhiker slung open the door and leaped out before we stopped moving. Mason had reached out to brace himself on the dashboard, but changed his mind suddenly and got his head cracked good. I reached out with one arm to keep him from sliding out the door into the ditch. The hitchhiker grabbed his other arm to drag Mason out with him. Our eyes met for a second. I didn't look at the gun, but I knew it was pointed my way. Damned if he was going to get Mason out of that truck to use as a shield. Not if I could help it.

He didn't shoot, I think he didn't want to waste the time or the bullets. He must have known he'd need all he had left of both. Already I heard the police shout a warning, heard the first crack of a pistol like a huge firecracker. I hauled at Mason's arm, determined to keep him inside—the hitchhiker let go and gave me a funny little cat grin before he disappeared back of the truck. All that happened in about three split seconds. Then Mason jerked me down across the seat with him.

"You want to get shot?" he whispered hoarsely. I hadn't even thought about the police accidentally shooting us. The window of the open door shattered and disappeared. The firecrackers were bursting in uneven strings now, they didn't sound anything like hunting shots. We hugged the seat covers, and I thought this was probably what it was like to be in a war, wondering if I had a gun, would I climb out and join in.

* * *

We lay there piled up, just moving enough to breathe, even after it was quiet.

"You kids all right?" A policeman peered in the window.

We sat up slowly, and after a second we climbed out the door. There were about three highway patrol cars there, and two more cars that must have been unmarked cop cars, because they were full of police. The police radios were scratching and spitting like a den of angry cats, sirens were going, the police were yelling back and forth at each other. But after the shooting, it almost seemed quiet.

Mason walked to the back of the truck and hung onto the sides of the truckbed, resting his head on his wrists, gasping.

"He was holding a gun on us," I said to the policeman, keeping an eye on Mason. He was in bad shape. And now that it was all over, I was a little shaky myself. "He was goin' to make us drive him to the state line."

The cop nodded. "There's been an APB out on him. Somebody spotted you guys picking him up and called us. He escaped McAllister early yesterday, killed a trustee while he was at it. Then he got to the city and shot a guy there. That one's going to make it, though. That was a real risky stunt you pulled, kid. That punk would have killed you without batting an eyelash."

He wasn't telling me anything I didn't already

know, but hearing it made it scarier. "I would just as soon have witnesses," I said.

"Well, you're real brave, real stupid, or real lucky," said the cop.

"Yeah, I been hearing that all my life," I told him.

"Oh, geez," Mason groaned. He had straightened up and looked around, just as a couple of highway patrol guys moved away and you could see the hitchhiker laying there on the brown grass, not four feet away. There was a red stain on his Levi shirt and a bigger stain spread on the grass under him.

I started toward him.

"Tex." Mason reached out to stop me. I shook him off and went on.

His sunglasses had fallen off but nobody had shut his eyes yet. He was staring up at the sky he couldn't see anymore with a bitter expression in his strange-colored eyes. I felt funny looking at him, almost like crying. He wouldn't be dead if it wasn't for me. Then I pictured Mason laying there, just like that. I didn't feel like crying anymore. I walked back to the truck.

Mason was talking to a policeman, but he broke off long enough to say, "Did you have to go take a gawk?"

"Yeah," I said, "I did."

I think we were there at least an hour more, telling the same thing over and over again, answering the same questions, talking to a dozen different people, including a couple of newscasters that showed up out of nowhere. Any other time I'd have been really excited about being on television, but it didn't seem important now. Mason stayed cool. But he had to work at it.

Somebody kept trying to get us to go to the hospital. I couldn't figure out why. Mason had a bad-looking bruise on his forehead—when I asked why he'd let go of the dashboard at the last minute, he said, "I'm not about to break my arm, going into the season." That was what I'd call basketball fanaticism.

I had a sore chest from getting thrown into the steering wheel; I'd probably be really sore tomorrow, but neither one of us was hospital cases. They seemed to think we should go because we'd had a bad scare. I never knew you could end up in a hospital from being scared.

The ambulance came for the body, and gradually the police started clearing away. The traffic that had been blocked up was moving again, slowing down as it went past like we were some kind of circus.

Mason had quit shaking except for a little shiver once in a while. I was beginning to think he'd be okay after all, when he asked one of the policemen if he knew what the hitchhiker had been in prison for.

"Assault, last time. He'd been in before. Hey, Ralph, you know what Jennings went into the can for the first time?"

"Yeah," the other cop came over. "A friend of mine busted him the first time. Drugs."

Mason went as white as a sheet, walked to the front of the truck, and threw up. The policeman told me it was delayed shock. I agreed with him, since the alternative would be telling him about Lem.

It was way after dark when we finally got headed home.

"I'm glad of one thing," I said. "They didn't ask to see my driver's license."

Mason half-laughed and half-sighed. "Texas," he said, "why did you have to go look, after they'd killed him? It wasn't exactly a side show at the Fair."

I was shocked that he could think such a thing. What kind of a creep did he think I was, anyway?

"I had to," I said finally. "Mason, I killed that guy, as sure as if I'd pulled the trigger. I knew it when I ditched the truck. I couldn't just walk off like nothing had happened. I had to face what I did."

"You're not sorry?"

"Well, I ain't sorry I'm alive, or sorry you're alive, and I figure that was my choice. Mace, something really bad must have happened to that guy. I mean, he was really a terrible person."

Mason just nodded. It was a little bit later that he surprised me by saying, "You don't think you could ever turn out like that?"

I thought a little bit before I answered—it sure was a time for thinking about things.

"Well, I don't think so. But then nothing really bad has ever happened to me."

"That's true," Mason said carefully. It was then I knew who it was that guy had reminded me of.

It was me.

SEVEN

I called Johnny as soon as we got home, to see if he could come over and watch the news with us. I wouldn't tell him why. He said he couldn't get out of the house, but he'd watch the news. I think he got the idea that I was going to be on it.

"I don't see what the big deal is," Mason said, as I turned on the TV and sat as close to it as possible. He was really working hard at staying cool. "It's not like you're starring in your own series or something."

He was just as excited as I was, but he was knocking himself out not to show it. I'm glad I'm not like that.

He sat down with me about one minute before ten. "I hope they didn't get too much of the crummy truck in the film." Suddenly he paused. "You think they're going to show that interview you did with that lady reporter? If they do, man, I'm going to kill you . . ."

I was hoping he'd forgotten about that. The lady reporter had asked me the same questions as the other ones, like, "Were you scared? How do you feel now? Did you think he would have killed you?" Stuff like that, then out of the clear blue sky she said, "Where on earth did you get those dimples?"

And I said nervously, "God gave me my face. But He let me pick my nose."

I thought Mason would have a fit! And if they put it on television . . .

The first of the news was just the usual foreign countries fighting. Then the newscaster said, "Here on the local scene an attempted kidnapping left an escaped convict dead and two area teen-agers shaken . . ."

I couldn't hear the rest of it. We were on television! There was me and Mason and the truck half in the ditch and the police standing around and a quick shot of the body being loaded into the ambulance. Then the reporters asking us questions—I looked younger than I thought I did. Mason came off looking pretty cool, he sounded a lot calmer than he had been. My voice sounded funny. I didn't realize I had a drawl like that. There was a close-up of both of us, then it was over.

"Hey," I said, "that wasn't very long."

"It was long enough," Mason said.

"Yeah, but the reporters were out there a

long time just to get a couple of minutes of film. You'd think they'd want it to be a little longer."

"Kidnappings are a dime a dozen," Mace said. I could tell he was relieved. We didn't look as awful as he thought we would.

"They didn't say anything about you bein' a basketball player," I said, "too bad."

As soon as the words were out of my mouth, the sportscaster came on and said, "One of the two teenagers involved in the kidnapping was Mason McCormick, the Garyville basketball star. Fans will be relieved to know Mason was uninjured and will be ready for the upcoming season."

"Huh," I said, looking at Mason. "I guess you liked that all right."

He did his best to look cool. "I guess so."

I threw a sofa pillow at him, but missed because he got up to answer the phone.

We were famous. Everybody we knew and a lot of people we didn't kept calling to ask us about being kidnapped. It took about five minutes of this to drive Mason up the wall. After that I answered the phone. After two hours I was so hoarse I couldn't talk anymore. So we took the phone off the hook for a while.

I felt real funny. I'd had a long day and it was after midnight and I could tell I was tired, but I was so wound up I could actually feel the nerves humming along in my body like charged electric wires. Mason, who would usually kill for his eight hours of sleep, sat around and drummed his fingers, or got up and paced around, like a chained dog.

"Tex," he said finally. He looked kind of sheepish. "You still got that joint Lem gave you awhile back?"

"Yeah," I whispered. I went and got it out from under the mattress. I handed it to him. Mason looked at it a second, then went to light it on the kitchen burner. He came back in and sat next to me.

"Lem would love to see this," I croaked. I wondered if I was ever going to be able to talk again. Mace leaned back and closed his eyes. "Yeah, I just bet he would."

I inhaled, held it, and passed the joint back to Mason. Grass isn't my favorite high, but I'll say this for the stuff, it really puts me to sleep.

Mason choked a little on his next hit and wiped his eyes. "I'm going to have to smoke myself silly before I quit feeling that gun in my ribs."

"Well, that'll only take a couple of puffs," I said. We started laughing. It wasn't that funny, but you know how you get when you're doing dope. I think that's the first time me and Mace ever just sat down together and got stoned. It was a weird ending to a weird day.

I almost went to sleep on the couch. I watched Mace walk across the room and drop the phone back on the receiver. It looked like he was moving in slow motion.

"I'm going to bed," he said. I nodded, unable to get up. The phone started ringing again.

"Oh hell," Mason's voice was squeaky from the grass. "I could do without this famous stuff."

He answered the phone and even across the room I recognized the voice that said, "What have you been doing? It's two in the morning."

Mason looked at the phone in his hand for a second, like he couldn't believe what he was hearing. Then he said, "What have *you* been doing? It's the first of November!"

It was Pop! I stumbled across the room and tried to get the phone away from Mason, who kept brushing me off like a pestering fly. I was trying to shake the fuzz out of my ears. That damn grass had left my head so spacy it took a minute to hear what Mason was saying, trying to sound normal. "Well, gee wheez, if I'd known all it'd take to get you to come home was getting kidnapped and almost murdered and four-state news coverage I'd have arranged it a long time ago. I mean, wow, something as easy as that—"

His voice broke off. He was white and his hand was shaking till the phone was hammering against his ear. I grabbed the phone away.

"It's me now, Pop."

Mason dropped onto the sofa and sat there staring straight ahead, one hand clutching the other in a white-knuckled grip. I watched him while I kept on talking.

"Yeah, yeah, he's okay, he's just sh-shook up, you know we had a rough day. You saw us on the news? In Dallas? Wow. Yeah, we're okay, don't pay any attention to that, Pop, you know how Mace gets when he's nervous . . . listen, we found out today that he's got an ulcer and—"

Mason made an angry gesture for me to shut up, but I kept on going. "Yeah, an ulcer, he's got to quit eating some stuff and calm down some—yeah, I think it'd help if you were here. Tomorrow? Great. Yeah, he's still here. Mace?" I held out the phone, but Mason shook his head.

"Well, Pop, he's kind of sick or something right now. I think his stomach is bothering him . . . sure, I'll tell him. See you tomorrow. Bye." I danced around the room, whooping. Then I stopped.

"What's wrong with you? Aren't you glad he's coming home?"

"Who cares?" Mason said. "It's not any big deal."

His voice was shaking and if it'd been anybody else besides Mason, I'd have sworn he was about to start crying.

"We just happened to cross his mind because he saw us on the news. Just a little reminder—"

His voice broke off. I sat down next to him. He was really upset.

"Hey," I said, "it's going to be okay now. Everything is going to be great."

He just shook his head. "It's not going to be the big deal you think it is. You'll see. You'll see."

I put my arm across his shoulder and patted

him, and he was stoned enough to let me. Poor Mace had been through a lot that day, what with his ulcer and Lem and that hitchhiker and being on the news. So I didn't argue with him about Pop. He was wrong, though. I was pretty sure he was wrong.

"I hear a car coming." I went to the front door to look for about the two hundredth time that day.

"Even if he left at six in the morning, which isn't likely, he couldn't be here this quick," Mason told me. He was making a big show out of being unconcerned. He was driving me nuts.

"You don't believe he's going to show up, do you?" I said, still watching the door.

"Maybe he will, maybe he won't. Who cares?"

"Man, you were on his case because he didn't come home, and now you don't care if he does. Make up your mind, willya?"

Mason didn't say anything.

"Mason," I said, "I thought you said it wasn't likely that he left at six this morning . . ."

Mason looked at me. I was laughing at him. He fought with himself for about two seconds before he jumped up and tried to beat me out the door.

Pop barely had time to get out of the station wagon before we were all over him. By the time we all got through hugging and dancing around and laughing, we were back in the house. Pop stood still for a second and looked us over.

"Boy, have you two grown!" He seemed slightly shocked. Mason was taller than he was by a couple of inches. Then he said, "Either that or I've shrunk."

We all laughed. But all of a sudden it hit me that Pop was a completely separate person from us. I don't know exactly how to explain it. He was just the same as he always was, but he was unconnected. Almost like he was a visitor.

We had barely got him unpacked and the major news rehashed when Mason asked, "How long you going to stay this time?"

It was like he was mad at himself for being glad Pop was home. I figured he'd start something like that, but wasn't expecting it so soon. I felt like slugging him.

"Mace, I don't blame you for bein' irritated with me. I was shocked myself when I realized how long I was gone. A month seems like a week used to. Time sure is getting screwy, the older I get . . . anyway, I quit rodeoing. For good."

"No kiddin'?" I said. Mason said, "I'll believe it when I see it."

"Honest, kid, I quit last spring. Then a guy I knew talked me into going to New Mexico to mine for uranium—that was where I was this summer. I'm too old for rodeo—have been for years, but you know, I loved it. It's changed a lot, what with the team leagues and the rodeo schools. I never expected to make a fortune at it, but now even having a good time costs money."

"Having kids costs money," Mason said. Pop was quiet, studying his hands where they rested on the kitchen table. They were brown and calloused and sprinkled with liver-colored freckles, sturdy and square-fingered, completely different from Mason's. I'd heard one of Pop's friends joking with him once, asking him how an old quarterhorse like him had sired a couple of lanky racers, and Pop said, "Their momma was a thoroughbred."

"Mace," he said now, "you're mad at me, and I don't blame you, but I'd like to hear it all now and get it over with, so we can go on. You got a bad tendency to bottle things up and dwell on them and brood about them, and I'd rather have a big explosion and then the clear air."

"Okay. All right. I've been known to explode. Like a couple of months ago when I hadn't heard from you since spring and the money was gone and the gas got shut off and I had to sell the horses since it was all I could do to feed us."

Pop looked blank. "You sold your horse?"

"Yeah, I did, and Negrito, too. And when Tex got mad about it I beat him up. Look at that scar. He's going to have it the rest of his life."

I hadn't realized Mason felt so guilty about that. The scar didn't bother me none. But whenever I let myself think about Negrito, it was all I could do to keep from getting into that fight all over again.

"I'm sorry, Mace, I never thought about the money—when I left I thought I'd left a good-sized hunk in the bank, and both you kids had jobs and, anyway, I didn't plan on bein' gone this long."

Mason was trying to keep a grip on himself, but I wouldn't have been surprised to see him break out in a foam like a frenzied horse. Neither one of them paid any attention to me. I always felt left out when they fought.

"I know you never thought about the money. The good-sized hunk shrunk real quick when Tex fractured his arm last May and we didn't have any insurance. We both had jobs, yeah. Summer jobs. It hasn't been summer for a while now. And whether you planned it or not, you've been gone this long. And I'd like to know how the hell long you're going to stay."

"You're sure not giving him any reason to stay," I said.

"You—you just shut up," Mason said. He didn't feel so bad about that scar that he wasn't ready to give me another one.

"Well," Pop said quietly, "it won't do any good now to say I'm sorry. I sure didn't think about things being that bad for you. All I can do is try to prove myself this time, and all you can do is give me the chance." That's okay, I thought. We can do that.

Mason looked away, drumming his fingers on the table. There was a long silence. "I guess it is all I can do," he said finally.

Pop watched him wistfully. Mason's good opinion meant a lot to him. His face brightened. "I tell you what, let's start off by getting those horses back."

"Whaa hoo!" I shouted, jumping up and turning over the chair.

Mason shook his head. "I don't want mine back, I'm not going to be around here a lot longer and I don't have time to take care of him, anyway. But you could get Negrito back for Tex."

"Sure," Pop said. "I got a little money saved up. I tell you what, Texas, I need a couple of weeks to get a job, and you got a birthday coming up then, so can you wait till the end of the month?"

I stopped dancing around, stared at Pop for a second, then choked on a laugh. I couldn't stop laughing.

"What's so funny?" Pop asked. I shook my head, trying to get my breath.

"October," Mason said. His voice was like a knife of ice.

"What?" Pop said. Mason was absolutely white. It almost looked like he was suddenly scared. It sobered me up to see him.

"His birthday was the twenty-second of October. Last month."

"Oh. I was thinking it was November."

Mason didn't say a word. He just got up and left. Pretty soon we heard the pickup squealing out of the driveway.

Pop shook his head. "That young-un can get his back up over the silliest things. I'm sorry about that, Tex."

"Shoot," I said, "it doesn't bother me. It proves you didn't forget about it completely, which was what Mason tried to tell me. Anyway, I thought it was funny."

"I don't think Mace thought it was funny," Pop sighed. I had to agree with him there.

The light blinded me. I hung onto something next to me to keep from falling into the white space. I heard the voice again and this time I could make out the words:

"What's goin' on?"

It was Pop. And I heard Mason's voice saying roughly, "Nothing is going on. He's having a nightmare, that's all."

My eyes adjusted to the light. It was just the bedroom, not some big expanse of empty white space. I was sitting on the edge of the bed, like I was ready to get up and go somewhere. I was in a cold, sick sweat. I slowly realized I was clutching Mason's arm in a grip that'd leave bruises, and I tried to let go.

"I'm okay," I whispered.

"Gosh, Tex, do you still have nightmares? I thought you'd outgrown that."

I shook my head, still unable to talk too much.

"We were kidnapped at gunpoint not too long ago," Mason said. "I reckon that could give anybody nightmares. I've dreamed about it myself."

"Sorry I woke y'all up," I said, trying to convince myself I really was awake, that the terror was over. "I'm okay now."

I dug my fingers out of Mason's arm and tried my best to look okay. I couldn't quit shivering. Pop look dubious. "You sure?"

"Yeah. I didn't mean to be hollerin'."

"Well, maybe we can all get back to sleep now. You guys have to be at school pretty early." Pop switched off the light and went back to his sleeping bag on the couch.

I swung myself back under the quilts. Mason crawled back around to his side.

"Mace?" I said. "You really have nightmares about that hitchhiker?"

"Yeah," he said. "I figure they'll go away pretty soon."

I was quiet. "I don't think that's what I was dreamin' about," I said finally.

"I didn't figure it was," he said.

I never did dream about the hitchhiker, and what's more, I didn't think about him much. All that seemed unimportant now that Pop was back, and Negrito was coming home. I did kind of get a kick out of the fuss people made over me at school, though. Nobody else in the whole school had ever been kidnapped, and only one other kid had been on the news, and that had been in grade school, in a spelling bee.

Miss Carlson was absent that day and we had a substitute teacher. We did all the usual things we do to substitutes, coughing at exactly one minute till, forming lines at the pencil sharpener, till she slapped a pop quiz on us. I got the feeling she'd been a substitute teacher before.

Before class was over she sent me to the office for talking. It wasn't my fault, really. Everybody wanted to know about the hitchhiker. Fortunately Mrs. Johnson saw my side of it, just told me not to let fame go to my head. While I was in the office I heard somebody say Miss Carlson had gone to a funeral, and then somebody else said it was the hitchhiker's funeral.

That bothered me. It took the fun out of being famous. I never thought about him having a funeral, or somebody going to it if he did. I hadn't thought about anybody missing him.

When Miss Carlson showed up the next day, I decided to find out for sure.

"Uh, Miss Carlson," I said, standing at her desk after everybody else had gone on to their next class, "somebody told me you went to that guy's funereal, the one the highway patrol shot."

"Yes," she said. "I did."

She didn't look like she was mad at me about it. She had real long eyelashes. I bet she was good-looking when she was young.

"Was he a relative or something?" That was what I was afraid of.

"No. Not even a friend, really." She paused, like she was hunting for the right words. Finally she said, "I read a book once that ended

TEX   139

with the words 'the incommunicable past.' You can only share the past with someone who's shared it with you. So I can't explain to you what Mark was to me, exactly. I knew him a long time ago."

I stood there, feeling like I do when I bump into things, not knowing what to do. "I'm sorry."

Miss Carlson shook her head. "Tex, please don't let it worry you. I'm sad about what happened, but not surprised." She glanced down into her grade book. "Now what ever happened to that other book report?"

I couldn't wait till the end of the month. Negrito would be coming home! Pop didn't get his old job back, at the cement plant, but he got another one, at the feed mill, pretty quick. Pop never had much trouble getting jobs. People tend to like him.

The day he was due home with his first two-week paycheck, I went bouncing through school like a ricocheting bullet. Somehow I didn't get sent to the office, though. Johnny broke the speed limit getting me home—riding double on his machine was breaking the law anyway, so it didn't take much to get him to break two.

Then it was an hour to wait till Pop got home. I thought I was going to go nuts. I went up to the horse pen and straightened up a couple of sagging fence posts and tacked up a strand of loose barbed wire. I'd spent the week before putting the rails back up on the pickup, so we'd have something to cart him home in. Negrito loaded surprisingly easy for a high-strung horse. All you had to do was show him a bucket of grain in the truck.

Pop wasn't home an hour later. Another hour later he still wasn't home. I got to thinking he had a car wreck. Mason came in, hot and sweaty from jogging.

"No, I don't think he had a wreck," Mason said, dropping into a kitchen chair and gulping buttermilk straight from the carton. "It just slipped his mind."

"Naw, something happened."

"Nothing happened except somebody probably asked him to stop off and have a beer."

About that time the phone rang. I let Mason get it. He got most of the phone calls at our house. A lot of them from girls.

"Oh, yeah? Well, what about going to get Tex's horse back? Forget about that?"

I went sick inside. Pop had forgot. Damn Mason, I got so tired of him being right all the time.

"Yeah, I know where he is. Sure, I'll do it. Will the check bounce? How much? Pop, I'll have to offer more than they paid me for him. Okay. Yeah, sure. Good luck."

Mason came back in the kitchen. "Me and you'll have to go after him. Pop's in a pool tournament over in Broken Arrow. No telling how long he'll be there. Says he's sorry, he just clean forgot."

"Well, at least he's giving me the money to get him back," I said defensively.

"I thought I was supposed to be the stubborn one in this family," Mason said.

My disappointment was beginning to fade. I was getting excited again. "Well come on, let's go!"

"No way. I got to take a bath first.

"Mason!"

"It'll just take ten minutes."

I started swearing at him, but he went on to the bathroom anyway. I had to resist an urge to go hold his head under water.

We got on the road finally. We had to drive clear to Muskogee.

"Boy, you really made sure I couldn't find him again, didn't you?" I said.

"That would have been all I needed, you getting arrested for horse stealing."

"You're the one that should have been arrested," I said.

Mason didn't say anything. We drove through Muskogee and turned down a blacktop road. It was getting dark, but you could see we were driving through a little housing development. The houses were each set on a couple of acres of land.

"You know your way around here pretty well," I remarked.

"I came out to look the place over first. I told you I made sure those horses got good homes, didn't I?"

Him and his truth hang-up, I thought sourly. Then I brightened up as we turned into a driveway. I heard hoofbeats as soon as we got out. "I'll be around back!" I shouted. Mason went to the front door. He could take care of the business end of it. I wanted to see Negrito.

There were floodlights turned on over a small wood corral in back. Negrito was tearing around some barrels, set up for barrel racing. Even though he could turn on a dime and hand you back a nickel change, I had never done barrel racing with him and was amazed to see how good he'd caught onto it in so short a time. After bending around the last barrel so sharp his rider's foot nearly touched the ground, Negrito flattened out in a dead gallop finish. When she pulled him up, he was blowing through his nose and snorting, the way he did when he was happy. I decided I could take up barrel racing if he liked it so much.

His rider saw me. She nudged Negrito into a canter and had him do a sliding stop at the fence where I stood.

"Who are you?" she asked. I didn't even look at her. Negrito was so surprised to see me that his ears were practically touching and he kept nickering from way down in his chest. I had all I could do to keep from grabbing him around the neck and crying.

"He used to be my horse," I said. I reached out and stroked his neck. Man, he was clean. He must have been brushed morning noon and night to be that clean. It's a mess trying to get dust out of a winter coat.

"You're not the boy we bought him from," the girl said. Her voice sounded stiff. For the first time I looked at her, seeing her. She was about twelve or thirteen, blond, freckled, braced, her eyes a light sky color from behind her glasses.

"That was my brother. I didn't know he was selling him. It was sort of an accident."

The girl slid off and stood by Negrito's head, holding the reins tight, like she thought I might grab them away from her.

"We paid for him," she said. "It was fair and square."

"Sure," I said. Negrito was nibbling on my sleeve, the way he would just before biting a hunk out of you. I was getting so sick I couldn't see good. They weren't going to sell him. I wasn't going to get him back. Like she knew what I was thinking, the girl said, "He's my horse now."

I looked around at the nice little paddock, with an open-faced barn. They didn't need the money. They could feed him through the winter. She wouldn't come home from school and find that paddock empty.

"I had a pony, but he died," she was saying. "I didn't think I'd ever want another horse. Nicky was a birthday present."

Didn't even get his name right, I thought bitterly. I asked Negrito how he liked it here.

"Oh, great, man, great." His head bobbed up and down. "Good food, good fun, lots of attention."

I didn't remind him I'd given him lots of attention, too. Horses are like real little kids. Now is what's important.

"He bites," I said to the girl, not looking at her, still patting Negrito's neck.

"I know."

He'd put on weight, but he'd been worked enough to turn it into hard muscle. His thick winter coat was like velvet.

"He spooks at things, too," I said.

"I know."

"He ain't really scared though, he's mostly just playing. I never hit him for it, you shouldn't hit a horse unless you really have to."

I turned to her. Her face was stiff and she kept wrapping the reins around her wrists. "I know."

I heard Mason honking for me. I knew he hadn't been able to make a deal. I leaned my head against Negrito's neck for a second. Horses really smell good.

"He missed you," the girl said suddenly. I looked at her and she seemed to be sorry she let me have that much, "at first."

I gave Negrito a final pat and turned away.

"I know," I said.

"He just wouldn't sell. I offered him more than they gave for him, but he didn't even listen. Said his kid was happy with the horse and he wasn't going to upset her. Seems like her pony died last year and he thought she never would get over it. I did what I could."

"Yeah," I said, hardly able to talk for the ache in my throat. "You did what you could all right."

Mason got a defensive look on his face. "Well, it's a good home."

"He had a good home."

Mason didn't say anything, stepping on the speed a little when we passed a hitchhiker.

"Mace," I said evenly, trying to keep my voice from shaking, "I am going to hate you the rest of my life for this. I mean it."

Mason looked straight down the highway. "Who cares?" he said. But I'd seen a muscle in his jaw jump, and I knew I'd hurt him. It felt good.

It was the first time I realized hurting somebody could feel really good.

EIGHT

Somehow, losing Negrito that second time was harder than the first time. It was just knowing he really was gone for good, somebody else was feeding him and brushing him and he was watching for somebody else in the mornings and after school that made me feel like I had a constant toothache or something. Even when I was thinking about something else, I could feel it in the back of my mind. The only thing that could really take my mind off Negrito was Jamie.

We were kind of going together. I couldn't figure out exactly how it had happened, except that we started meeting between classes and had lunch together and I quit riding home on Johnny's cycle so me and Jamie could go to the drugstore and get a Coke before we rode the bus home. Everybody in the school knew we were going together. I used to wonder how guys ever got the nerve to ask a girl for a date, but since Jamie had been my friend before she was my girlfriend, it was really easy to say at lunch one day, "You think Cole'd let you go out with me?"

She shook her head. Her hair curved around her face like dark feathers.

For a second there I hated Cole Collins. Then I didn't hate him, because he was Jamie's father and I'd have to learn to get along with him.

"It's not just you," she said. "Cole thinks I'm too young to date anybody. No car dates until I'm sixteen. And Mona agrees with him. All she ever does is agree with him. If I ever get married, I'm never going to agree with my husband."

I raised my eyebrows. "Not ever?"

She looked at me with the eyes of a wicked colt. "Oh, maybe sometimes . . . Anyway, I bet Cole wouldn't mind me going to watch Bob play basketball. If you were watching Mason, we could sit together."

"Cole doesn't go to the games?" I asked. Pop never missed one.

"Cole went through all that stuff with Charlie, only it was football. And when Blackie refused to go out for anything, sports got to be a family hassle. Cole and Blackie really had some go-rounds . . . you know what? Cole would like to have Mason for a kid, I bet. And Mason would like a father like Cole."

"Huh," I said, because I figured it was politer than saying, "Mace hates Cole's guts."

"No, really, Tex, listen. That time you and Johnny and Bob came home drunk, Cole kind of hinted real strong for us to stop hanging around with you guys, but I could tell he'd been impressed with Mason. He's not too impressed with any of us. Charlie's too much of a playboy and now Bob's got it into his head that he wants to be a priest; Johnny's such a scatter-brain I think Cole'll be relieved if he just makes it to twenty-one. And me. The little lady. Cole has the hardest time understanding that I'm a person, just like the rest of his kids, that being a girl doesn't mean I'm going to be sweet and dainty and grow up to be a devoted little mother just like Mona. Geez, it gives me cold chills just to think of it . . ."

She scrunched her face up like she was hearing squeaky chalk across the blackboard.

"How about Blackie?" I asked. This was really interesting. I never thought about what parents would want out of a kid. I thought you just took what you got.

"Don't ever tell anybody I told you this." She dropped her voice. "Swear?"

"I swear."

"You know when Blackie moved out—he didn't just move, he ran away from home. It was after a big fight with Cole about not going to college. Not playing football was bad enough, but Blackie didn't even want to go to college. He didn't even want to go to art school. Said he had to know what he could teach himself first. Man, it drove Cole nuts to argue with him, because Blackie never argued back. You know how quiet he was. Sometimes I couldn't tell he was in the same room with me. He just stood there and let Cole get madder and madder. Then that night he took off. He wrote Mona from San Francisco to let her know he was okay. Tex, Cole and Mona had some awful fights about it. I'd never heard them fight before. They didn't know we were listening. Me and Johnny sat on the stairs and listened to them and we both were crying like little kids . . ."

I reached over and took her hand. I couldn't stand the thought of her crying. She took a sudden deep breath. "Anyway, Cole hasn't made a big deal out of sports since. Basketball was Bob's idea. Blackie had the perfect build for football . . . it must be weird for him, to look like a football player and be totally different on the inside."

I had my mind on other things. "Mason sometimes goes to parties after the games," I said suddenly. "With a bunch of other people. So I could get the pickup and drive us around."

Suddenly I remembered Johnny. If I had the truck, he'd want to go driving around, too. It would be hard to tell him the truth. He'd never understand how I felt about Jamie. He was interested in girls, sure, but it was like being interested in *Playboy* pictures and stuff like that. He hadn't got to the point where he was interested in *real* girls. And even though he loved Jamie in the same way he loved their dog, he didn't quite realize that she was a girl, the kind of girl somebody would lay awake thinking about for hours.

"Johnny—" I began.

Jamie quirked the corners of her mouth down. "I'll tell Johnny he's not wanted."

"But . . ."

"Well, he's not, is he? All right. He'll take it a lot better from me than you."

Boy, she was mean. I really liked that. I really did.

They were out to kill Mason. It was plain from the second he stepped on the court and the opposing team started booing. People were trying to make him foul, or just plain knock him down and put him out of the game. They didn't know Mason. That kind of thing just made him cooler and cooler. He really played better when people were booing him than when our side cheered him. I took Mason pretty much for granted at home, but watching him on a basketball court kind of put you in awe. Man, he was good!

I don't think I could have stood having all those people not liking me. But then, Mason never cared much whether people liked him or not.

"Everybody does, though," I said to Jamie, after telling her that.

"You mean he's popular. Everybody thinks he's cool. Not everybody likes him."

I didn't want to know if that included her. It would really bother me if she didn't like Mason.

It took everybody screaming at once to get my mind back on the game. Everybody was on their feet screeching as Mason caught a rebound and made a wild shot that turned out to be a basket and on his way down from the leap an opposing player slammed into him. He came up off the floor so fast it looked like a bounce and for a second I thought that the other guy was going to get stuffed through the basket, head first. Whistles were blowing all over. It looked like everybody was going to rush into the middle of the court and start killing each other. People were just going crazy.

Mason stood there, holding his right elbow. As far away as I was I shivered. I remembered one other time I'd seen him fighting to control himself like that. He was right on the edge of blowing up like a stick of dynamite.

The coach came out and looked at Mason and then Mason turned and stalked off the court. The screaming was unbelievable. He sat down on the bench while the doctor looked at his elbow, bent his arm up and down some, then said something to Mason that made him shake his head angrily. You could see the doctor getting mad. Finally Mason got up and followed him out.

"Hey," I said, "he must be hurt."

I doubt that Jamie heard me. The guy who had knocked Mason down was being taken out of the game, boos and cheers following him. I looked over to where Pop was sitting with Ernie Driscoll's father. He was pushing through the crowd, going to the locker room.

"Come on," I grabbed Jamie's wrist. "I want to see how he is."

"You go. I've got a brother in this game, too, you know."

I didn't like leaving her there in that screaming mob, but on second thought, she could hold her own pretty well. I shoved my way off the bleachers and ran to the locker room.

"It's not serious," the doctor was saying.

"Then I can go back in," Mason said. His arm was in a sling. Pop watched him worriedly. Suddenly I remembered when I busted five ribs in a junior rodeo. Pop had been nice, concerned, but not real worried. Of course, my memory could be wrong. It's pretty good, though, mostly. But he always had worried about Mason more.

"Not this game. A lot of good you'd be with your right arm messed up," Pop said roughly.

Mason shrugged. "My left is just as good."

Mason is ambidextrous. That means he can use either his right or left hand. For some reason, when I was little, I thought that meant he was part water lizard. Don't ask me why.

"You're out of this game all right," the doctor said. "You're lucky you're not out of the season."

Just the thought of it made Mason flinch, and to hide it he said, "Hey, Tex. Want to trade clothes with me and go shoot a few baskets?"

"Sure," I said.

"Got another basketball player in the family?" the doctor asked.

"If we do, I didn't know about it," Pop said.

"Well, maybe you ought to take the time to find out," Mason snapped.

He never was much fun to be around if he was mad or hurting, so I just said, "If you're okay then I'm getting back to the game."

"Good idea. Go root for Bob Collins. He's gonna need all the help he can get, now."

Being a big shot didn't go to Mason's head. Much.

He was right, though. Poor Bob was everywhere, trying to make up for Mason not being anywhere. His main function had been to get the ball to Mason, and you could see Bob forget and pause and wonder where the heck Mason was. He even tried shooting baskets, which shows you how desperate he was: He wasn't too bad at it, though, for a short person.

Me and Jamie yelled ourselves hoarse for him, along with a few hundred other people, but it wasn't any use. We lost by six points.

"You'd think Mason was their lucky rabbit's foot or something!" Jamie griped as we drifted with the crowd out to the parking lot. Most of the people from our school were pretty mad. "They just gave up without him. I'd like to know just what it is that Mason's got. Look at Bob, smarter and nicer and twice as good-looking, and everybody likes him—but ask anybody who the most popular guy in the school is and they'll say 'Mace McCormick.' And Mason's too snotty to speak to half of them."

"Maybe being popular and being liked ain't the same thing." I said, deciding not to point out to her that the same things could be said about her. The bunch she ran with was the "popular" girl group, but they weren't all well-liked, even Jamie.

The Riverview people were acting pretty silly, jeering and cheering and stuff, but I didn't pay much attention till Ralph Hernesy poked me and said, "I guess that shows you, huh, Mac?"

Even though we went to different schools I'd known him a long time, from horse shows and rodeos. And school games.

"Oh, bug off," Jamie said. Or something like that.

Now I wasn't in the best mood I've ever been in. I probably wasn't as bad off as some of us were, judging from looks on faces, but losing the game, and knowing we only lost because they'd put a hit man on Mason, hadn't exactly made me cheerful.

Ralph looked at Jamie. "I never thought you was any judge of horses, Tex, and you sure ain't any judge of a . . ."

I belted him so quick I didn't realize what I was doing till after it was done and Ralph was sitting on the ground spitting out a tooth.

Somebody shoved me in the back. "What's the matter? Sore loser?"

I whirled around, but the shover took off and disappeared in the crowd. I turned back to Ralph. He was crawling around frantically, looking for his tooth. "That was my false tooth, dammit!"

"Ain't you a little young for false teeth?" I asked, dumbfounded.

"I knocked the real one out last year at a rodeo. My Mom will kill me! It cost a fortune."

I squatted down beside him, licking the blood off my skinned knuckles. "Huh. Can you put it back in if you find it?"

"I think so."

I looked around for it for a second, till Jamie poked me with her foot and said, "I think we're going to get to see a riot."

All round us people were shoving each other, or already into fights. I jumped up and grabbed Jamie's wrist, dragging her through the crowd.

"Wait! I want to see what's going to happen!"

I blocked a punch somebody threw at me and speeded up.

"Sweet stuff," I told her, "if you want to watch a riot, watch it on TV. If you're there, you're *in* it."

I opened the door to the pickup and shoved her in. You could hear the police sirens coming.

I didn't particularly want to renew my acquaintance with the town cops, so I laid a little rubber getting out of the parking lot.

Jamie twisted around on her knees to look out the cab window.

"I wanted to know what was going to happen!"

I slapped her on the bottom, and she turned back around and slid down onto the seat.

"You'll hear all about it tomorrow."

"Secondhand."

"Yeah. The black eyes'll be secondhand, too." I paused. "What time you supposed to meet Johnny and Bob?"

"In an hour. At the car wash. We've got time to get a Coke or something."

I followed the highway to the gravel pit road and turned off. After a mile or so I pulled the truck over and switched off the lights.

"They sell Cokes around here?" Jamie asked mildly. In the cold starlight her eyes glittered like a cat's.

"You had a Coke at the game and I ain't thirsty." I said. I put my arm around her. I had been thinking about this for a long time. I kissed her, soft, so not to spook her, but it wasn't any rinky-dinky Mickey Mouse kiss. After the first one, I wasn't fooling. I loved her so much it seemed like she was a part of me, or should be, or there was a way for her to be . . .

"Stop it!"

My heart was thumping in my ears so hard I barely heard her. She was wiggling in my arms like a landed bass; she got her hands on my chest and shoved. I let go, trembling from trying not to crush her. I couldn't tell how much time had passed—minutes or hours.

"What's wrong?" I asked, when I could get my breath.

She tugged her sweatshirt down. "I am not ready for this. I mean it."

I stared at her, completely mixed up. Looking at her without touching her was almost a real, physical pain. She must know how I feel, I thought, she wouldn't be that mean to me . . . I reached for her again.

"I mean it, Texas," she warned.

Man, that hurt me. I just didn't get it. I slid back to my side of the truck, gripping the steering wheel so hard my knuckles turned white.

"Well," I said, as soon as I thought I could talk okay. "What did you let me get started for?"

"I didn't know you'd be in such an all-fired hurry. Anyway, I was curious."

Curious. I was burning up and she was curious. Something was really wrong here.

"I hate it when Cole is right," Jamie said suddenly.

"What's Cole got to do with this?" I asked tiredly, resting my head against the steering wheel.

"He said I was too young to start dating. I mean, dating even. We haven't even gone to the movies yet."

"Okay. We can go to a movie." The last thing I wanted to do was go to a movie. I'd just be waiting for it to get over with so we could come here and make out. "Jamie, I love you."

"Look Tex, I love you too." She certainly sounded matter-of-fact about it. "Right now I think you're the only boy I'll ever feel this way about, but, then, I'm probably wrong about that. But even if you are—look, my life is complicated enough right now. Sometimes I think I hate everybody, and sometimes I think I love everybody, and sometimes I'm mean and hateful to people, like Johnny or Bob, just to see if I can hurt them, but I love them and I'm sorry, after. I get mad at Cole for not understanding anything and mad at Mona for understanding everything. A lot of times I can't stand the way I act. I mean, I know people think I'm a bitch. And then I think if people don't like the way I act they can go jump in the lake. Then I worry that nobody likes me. See? See? I'm having enough trouble figuring things out right now without throwing in sex."

When she said "sex" I felt my face go red. I know it sounds dumb, but I hadn't thought of what we were doing as sex.

"I guess you don't love me as much as I love you," I said.

"You turn pitiful on me and I won't love you a bit."

I laughed a little, even though I felt like spanking her.

"Jamie, when we get older, sixteen maybe, let's get married."

I knew I'd never feel this way about any other girl. I wanted to know Jamie was going to be there the rest of my life.

"I can see me marrying you," Jamie said slowly.

"Yeah?"

"Yeah. When I'm eighteen or nineteen and scared of the way things are changing, the way people are going off in different directions, and the simple life looks romantic, a good way to keep everything the same . . . yeah, I can see me marrying you. It'd last about a year."

If she'd thrown a bucket of cold water over me it wouldn't have done a better job of cooling me off. I was even shivering a little bit as I started the pickup. All the feeling had been wrung out of me.

"I forget," I said, making a U-turn on the dark road. "You're one of them that's going."

Johnny and Bob were waiting at the car wash, sitting in Denny Brogan's car.

"See you at lunch Monday?" Jamie asked, before she opened her door. I'd been quiet all the way back and she was getting uneasy about it. I shrugged. "I don't care."

I did, though. I really did.

"Well, neither do I!" She jumped out of the truck and slammed the door so hard it cracked the window right down the middle. I just looked at it. I'd seen Mason do the same thing twice before. Bob took off in a big hurry—I guess he was late for a date. I just sat there, watching people drive up and down, talking to whoever pulled over and got in the truck with me. Everybody hung out at the car wash on Saturday night, sitting on their cars, driving by, looking for booze or dope or just company. I was just out for some company.

You could pick up girls there, too, but if I couldn't have Jamie, I didn't want anybody.

NINE

"You think this is going to be worth it?" Johnny whispered. "What if they find out who did it?"

"Nobody's going to find out who did it," I whispered back. We were gluing individual caps on the keys of the typewriters in office machines class. As soon as the key hit the paper, the cap would explode. Since this was the day of the nine-week test, there would be a lot of keys hitting at the same time. There was always something depressing about a test day, anyway. I figured this little job might liven things up some. The month that had passed since my fight with Jamie had been really draggy.

Johnny and I got to school real early and used a special key Roger Genet loaned me. It would open about anything if you knew how to use it. I didn't ask Roger what he used it for.

"Sure it's worth it," I went on. "Even if they do find out who did it, what's going to happen? We'd get sent to the office, get a lecture and a couple of swats. That's nothing."

Johnny gave me a dry look. "Cole Collins isn't your father."

"Oh," I said. "Yeah. Well, anyway, nobody'll know it was us."

When we got through, we locked the door carefully and went out to the smoke hole, which was the road corner of the baseball field. Nobody else was there, it was still too early. Johnny had started smoking lately. I got the feeling it was to put something over on Cole. I've been meaning to take it up myself, but haven't got around to it yet.

"I guess Mason's all excited about his scholarship," Johnny said.

"You'd think so, having his pick of a few like that," I answered. "But he's still so strung out sometimes I think he's lost his marbles. I figured once basketball was over and he was sure he was going to college he'd calm down some. But he's on my back all the time, worse than before."

Sometimes I thought Mason hollered at me all the time to make up for Pop not hollering at all. It was like he was constantly poking and prodding at Pop to make him do something—what, exactly, I didn't know, and Pop sure didn't, either. Sometimes he'd look at Mason like a chicken that had hatched a goose egg. If Mason was worried that Pop wasn't paying enough attention to me, he could have saved himself the trouble. Mason would be gone for college pretty soon and then Pop would have to notice me a little more. I mean, I'd be the only kid, then.

"At least he's got a job now," Johnny was saying. He dropped his cigarette and ground it under his heel. It did look cool. I'll have to get around to smoking one of these days.

"Yeah, at least he's gone more. I used to think I was going to really miss ol' Mace when he left, but now I think I'll cheer all the way to the airport."

"Uh, you haven't seen Jamie lately, have you?" Johnny said. He sounded like he had rehearsed it.

Well, it had been one month and four days since the basketball game and I'd thought about her at least every hour since then, but I just said, "Not lately."

I'd see her in the halls and my heart would spook and take off at a pounding gallop. Then she'd just say, "Oh, hi," and I'd nod back, cool as possible. After that I'd want to either run up and hug her, or belt her. Or go off somewhere and cry. I didn't do any of them.

I wasn't about to say all that to Johnny.

"Any particular reason?" he asked.

For a second I wondered if Jamie had put

him up to this. Maybe she missed me as much as I did her. Then I thought, No way. She didn't care.

"No special reason," I lied calmly.

Johnny looked at me skeptically. "Geez, Tex, you still do like her, don't you?"

He sounded like he couldn't imagine why anybody'd like Jamie.

Me and Johnny could always talk about anything. That was what best friends were for. It was weird, not being able to say anything about this to him. I couldn't do it, though. Finally I just said, "Let's get back. We don't want to miss the fireworks."

"I've called your parents," Mrs. Johnson said.

Capping the typewriters had gone even better than we'd hoped. The dead silence of a school on nine-week test day; everybody a little tense, whether they cared about grades or not—test day'll do that to you; me and Johnny looking at each other, cracking up before anything happened; Miss Carlson frowning, "What are you two . . ."

And from the typing room next door, noise like a machine gun. Followed by shrieks.

Mrs. Bennett had to go home with a case of nerves. Miss Carlson couldn't get the class calmed down soon enough to have time for the whole test, so she had to divide it into two parts. We heard later that the news made it all over the school in one hour, and over to the high school the next. All in all, it went better than we ever hoped, except that we were the suspects and laughing too hard to deny it.

"Your fathers, I should say."

I stared down at the paperweight on Mrs. Johnson's desk, one of those balls you turn upside down to make it look like a snow scene. I tried to look sorry.

"You called my father? At work?" Johnny's face went white.

"The last time you got into trouble, your father asked me to call him if something happened again. Immediately. He'll be here soon. Tex, your father will be here after school. You can wait in the office until he gets here."

"All day?" I looked at the clock. It was 10:30. "What about the rest of the tests I'm having today?"

"You'll have to worry about that later. I'm not going to give you a chance to disrupt the rest of the day. Just have a seat in the foyer. You'll be there awhile."

The thought of sitting around all day was making me sick. But I didn't feel anywhere near as bad as Johnny. He looked like a ghost. I really felt sorry for him.

He suddenly jumped a little, and quickly jammed his hand down his shirt pocket. "Here," he whispered, handing me his cigarettes.

"Holy cow," I breathed, taking them. That would be all he needed, Cole catching him with cigarettes on top of everything else.

"Quiet, you two," snapped an office worker. I stuffed the package in my pocket hastily, hearing heavy footsteps in the hall. Johnny looked like he was about to throw up. I glanced up into the doorway, cringing a little myself. It wasn't Cole, it was Mason.

"Hey," I said, "what are you doing here?"

"I could ask you the same thing," Mason said.

My stomach plunged. I hadn't been too afraid of facing Pop, but the look on Mason's face was giving me chills.

"Oh, God, Mace, get outta here," Johnny begged, "Cole's going to be—"

Cole walked in the door. I froze where I was slouched in my chair. Johnny looked like he wished he were dead. Only Mason seemed unaffected. He just nodded at Cole like he probably would if he met him in the dime store.

Cole looked at Johnny, then at me. "I might have known," he said.

I managed to pull my legs in under me and straighten up some. Something about the tone of his voice made me feel like I was lower than an earthworm.

"Might have known what?" Mason asked, not hotheaded like you'd expect, just reasonable.

Mrs. Johnson came to the door of her office. "Since I do have a conference room, may I suggest we hold our conference there, instead of here in the hallway? Hello Mason, I don't remember calling you to this meeting."

"I'm here anyway," Mason said. "I knew Pop couldn't get off work and you'd want to talk to somebody about Tex."

"So here you are. Well, come in, everybody."

Everybody waited for Johnny and me to unfold ourselves and troop in first, like criminals going to the execution.

"Might have known what?" Mason asked again, once the office door was shut behind him. Like there hadn't been any interruption.

He could look Cole in the eye. He'd grown that much. Cole stared back at him steadily with those dark blue eyes he'd branded every one of his kids with.

"I might have known that your brother was behind this. He usually is, whenever Johnny gets into trouble."

"Or Johnny gets Tex into trouble. I think it's fifty-fifty," Mason said. He sounded calm, completely in control of himself. Not flaked out like he had been lately.

Cole turned to me. "Tell the truth—"

Man, with Cole Collins towering over you, you told the truth!

"Was this your idea or Johnny's?"

I started to open my mouth, but Mason said, "I know this was Tex's idea, and I know where he got the idea. The point is, it isn't always his fault. Last spring when he and Johnny were messing with the shopping carts at the Safeway store, and Johnny ran Tex into the side of the store and fractured his arm, did I come storming over to your place, trying to get you to lock Johnny up? No. I just figured next time would be Tex's turn to do something stupid."

Cole didn't look convinced. He went back to Johnny. "Didn't I tell you that this friendship wasn't going to do you any good? I want you to promise me that you're going to end it, right here."

Suddenly I was so glad Cole didn't know I was in love with Jamie. For the first time I understood what Romeo and Juliet was about, even though I never have been able to read the play.

"No," Johnny said.

Cole said "What?"

Johnny had more guts than I'll ever give myself credit for—he answered, "No, sir."

"Tex," Mason said suddenly, "tell the truth—" he was half-mocking Cole and didn't care if he did know it—"Are those your cigarettes in your shirt pocket?"

I gave him the dirtiest look I could come up with. Johnny was close to shaking.

"Yeah," I said, "they are."

Cole looked at me, and then at Johnny. Then he said to Mason, "Get to the point."

"The point is, neither one of these two turkeys is perfect. Both of them have a bad tendency toward trouble. But you can't blame Tex every time. He's not a bad kid, and he's not a bad influence, any more than Johnny is."

Suddenly I realized Mace was controlling himself because he cared about what Cole thought of him.

He didn't care if Pop or I saw him lose his temper, but here he was, breaking his back to get Cole's respect. And you could tell by the way Cole looked at him, that he'd gotten it.

Jamie was right, I thought, how weird.

"You may be right," Cole said at last. Shook up as he was, Johnny couldn't help giving me a look of amazement. Somebody other than Cole be right?!

"Well, now that we've agreed that the blame is to be shared equally, maybe you'd like to hear what the punishment is?" Mrs. Johnson said.

"I'd like to hear what your punishment is," Cole said. "What I have in mind may be different."

"Three days suspension. The nine-week tests will be made up every day after school, a test a day. They will receive a grade lower than the grade they score. And I'm sending a recommendation over to the high school that Johnny and Tex be placed in separate classes next year."

"That sounds fair," Cole said. Then he said, "John, what did I tell you would happen the next time you got into trouble at school?"

"You're going to sell the cycle."

Man, I was so mad I couldn't see straight. They were really being big shots. And when Cole glanced at Mason, I said hotly, "He's already sold my horse, I don't think he can do much else to me."

"I'll probably think of something," Mason said mildly.

"I'm sure you will, you lousy son of a bitch," Johnny said.

Silence. Me and Johnny looked at each other. He'd said it for both of us. And to both of them. We shared a split second of triumph, before Cole took Johnny by the shoulders and marched him out of the office.

"I'll be back this afternoon," Mason said to Mrs. Johnson. "When Pop gets here."

"Mason, I understand your concern, but do you think that's necessary?"

"Oh, yeah," he said, "oh, yeah."

Then it was just me and Mrs. Johnson, looking at each other over her desk. She sighed. "Tex, you better take a seat in the foyer again. You're going to have a long day."

"I reckon so," I paused. "Listen, Mrs. Johnson, I really am sorry. I didn't think it was going to cause all this trouble."

Usually she would give me a wry grin and say, "Try not to let it happen again." But today she set her jaw and said, "Tex, listen to me. You had better start thinking."

I kind of cringed out of her office like a whipped pup. I hate to get people I like mad at me. But I can't seem to stop doing things that make people mad. It is really strange.

Nothing exciting happened in the office that morning, except when I accidentally tripped somebody. My legs are growing so fast it's hard for me to keep track of where the end of 'em are.

I got to leave for twenty minutes to get some lunch, but sitting around had killed my appetite, so I went out to the smoke hole instead of the cafeteria. I was getting congratulated on the best stunt of the year, when I noticed a blue car parked along the road next to the baseball field. I ducked out of the group and ran over.

"Hey, Lem!" I opened the car door and hopped in. "What are you doing out here?"

Lem looked like I'd just woke him up from a nap. "Hey, Tex, how's it goin'? Oh, I'm waiting around for Dwayne Kirkpatrick. I'm tryin' to talk him into letting me have some of that third generation home-grown he's got. Man, I could get a hundred bucks a lid for that stuff. I gotta little bit left, want to try it? One hit'll last you the rest of the day."

"For a hundred bucks a lid it ought to last you the rest of the month. Shoot Lem, nobody's got a hundred bucks to spend on grass."

"You'd be surprised, man. And this is dynamite stuff, killer weed."

"Well, if you're waiting for Dwayne you'll be here all day. He skipped school to go fishing."

Lem shook his head. "That turkey—he's got no sense when it comes to money. He'll make the grade in aggie school, though, 'cause he sure knows how to breed his weeds."

"How's the baby?" I asked. Lem'd been doing grass himself—his eyes were reddish and the car smelled so strong I thought I was going to get a contact high. You know, even though I don't smoke grass much, I really like the way it smells. I always connect it with friendly people.

"Oh, he's fine. You know, that kid really is smarter than the average baby. Connie looked it up in a book. He's real advanced for his age. Big, too."

I hadn't seen Lem in months, not since that time we went to the city to see about Mason's ulcer, but he didn't seem very excited about seeing me. When somebody's smoking it's hard to get them excited about anything except ice cream.

"Still like it in the city?"

"Yeah, there's always something going on. I miss havin' horses, though. I ain't had a chance to ride in a year. When you going to get another horse, kid?"

"I don't want another horse," I said, my stomach tightening. "Losing Negrito was like losing my best friend."

"Well, I know how that feels," Lem said.

"Why don't you come by this afternoon?" I said. "Mason'd be glad to see you." I didn't realize I was lying till I saw the look Lem gave me. I went quiet.

Then Lem said, "You know, I used to think all Mason wanted was money. But that ain't it, or he'd be in this business with me, 'cause I'm rolling in it right now. But what he really wants out of life is to be respectable. If that ain't a hell of a goal."

"Different people go different places," I said, getting out. I had to be heading back to the office. "Say hey to Connie and the baby for me."

Lem seemed to be thinking hard for a minute, then he said, "I'm making a delivery out here this afternoon. You want a good deal on white crosses?"

I shook my head and waved him off. Maybe I was just depressed about going back and sitting around the office, but I had a strong feeling that Lem didn't belong in the city. There are people who go places and people who stay and Lem should have stayed.

Mrs. Johnson called me into her office when I got back.

"Tex," she began, then stopped. She sniffed suspiciously.

"Have you been smoking grass?"

Damn. I should have let my clothes air out a little before coming back in. "No, ma'am," I said. "I wouldn't want this day to drag out any longer than it has to."

"I suppose that's true. Anyway, what I wanted to talk to you about was a job this summer. Mr. Kencaide of Kencaide Quarter Horses contacted the school, wanting to hire some kids for summer work. Would you be interested?"

"Sure!" I said eagerly.

"Mr. Kencaide wanted me to emphasize that he doesn't want a bronc buster, or people out to play cowboys and Indians. But for some reason I think you'd behave responsibly in a job like that. I don't know why I think that, except I did have one or two people you mowed lawns for last summer call me and say they were pleased with your work. Here's Mr. Kencaide's card, give him a call and say I recommended you. Please don't give me any reason to be sorry I did."

"No, I'll do a good job. I take horses serious."

"That may be what saves you. Tex, would being expelled bother you?"

I almost dropped the card. "Expelled?" I managed finally. "Yeah, it would."

"I was seriously considering it today. In fact I've seriously considered it several times during the last three years. I haven't, because I like you. But I like an orderly school even more, and if it comes down to a choice, you just might lose out. Understand?"

I nodded. Expelled—I didn't even know anybody who'd been expelled, except Paula Luiz, for jumping on her home-ec teacher with a rolling pin.

"It's not very long until school's out—but remember, it's never too late to be expelled."

Mrs. Johnson's eyebrows twisted together in the middle when she was saying something serious not very seriously.

"Now just take your seat again. Your father ought to be here about four."

About four. It was noon. I'd heard the phrase killing time, but now I knew what it meant. I felt like I had murdered a whole day. I could never hack an office job. You know how it feels when your foot goes to sleep? Well, by four I felt like my whole body had gone to sleep, including my brain.

Mason came stalking in, right at four. He didn't say much, but it was pretty clear that he'd used up his quota of self-control for the day.

"Come on in, Mason," Mrs. Johnson said. "You, too, Tex."

Almost everybody had gone home, except the sixth-hour gym guys. You could hear them out on the baseball field. Mrs. Johnson wanted to get us out of the way of the janitor, who was mopping the office.

"I don't know if the old man is going to show up or not," Mason said abruptly.

"He said he would when I talked to him," Mrs. Johnson said. "If I remember right, he came up here to talk to me about you a few times. Your big problem was fighting."

It took all Mason's politeness not to scowl at her. He certainly didn't want to remind me that he'd been in trouble before, too.

"Yeah," I said. "I've never been big on fighting."

"But you've been in here for everything else since the seventh grade," Mason said. He didn't get into fights much anymore. Nobody wanted to mess with him.

"Well, anyway, this is the first time Pop's had to come in and talk about it," I said defensively.

Mason gave me a sarcastic grin. "It's not the first time he's been asked. It'll just be the first time he's showed up for it."

Just about then Mrs. Seymore, from seventh-grade speech, stuck her head in the door and said, "Helen, come here and look. You won't believe what I've found in a locker . . ."

That stuff had been going on all day. I never realized how much running around the building was involved in an office job.

"I think I can trust you two to stay here."

"Sure," I said. Mason didn't say anything. I wouldn't have cared if he left. Instead he paced around the room, getting madder and madder.

"I don't know why you have to go poking your nose in," I remarked. I was thinking it was a wonder he had any insides left. He'd had to go to the doctor again, just a few weeks ago.

"I am poking my nose in because I don't want to see you get expelled," he said savagely.

I started. "You knew I might get expelled?"

"God yes—they were laying bets on it over at the high school."

"Huh," I said, amazed. I sure hadn't known what a big deal I was causing.

"You had just better wake up a little, man—" Mason began, when Pop walked in.

"Surprise, surprise," Mason said. Pop took one look at him and decided to ignore him. When Mace was like that, sometimes it worked. Sometimes it didn't.

"I hear you got into some trouble," Pop said. He'd come straight from work, he was covered with oat dust and burlap fuzz.

"Yeah," I answered. "I glued caps on the typewriter keys this morning. School started off with a bang."

"Well, I wonder where you got that idea." He was trying hard to keep from grinning. I had known all along that Pop wasn't going to think this was real serious—especially since he was the one who had told us about doing the same thing in high school. It probably never occurred to him that I'd try it, but it was a little late to make it the crime of the century.

I couldn't see what else he could do, besides

take it calmly, but Mason was absolutely enraged.

"Okay." He stalked around the room like a frenzied panther. "Okay, so you can't take Tex serious. So you can't give a damn about what happens to him. All right, I'm trying to live with that. Then think about me! For God's sake, how do you think I feel, seeing you being 'nice' to him, like you'd be 'nice' to a goddamn stray puppy! While I'm the one who has to look out for him and what's going to happen when I'm not here?"

Pop and I were both staring at him. I was ready to call in the straightjacket people.

"Geez, make it easier on me if nothing else! He is my brother even if he isn't your son!"

It was so quiet. Just the far-off sounds of the baseball game, and a wood-dove somewhere. It seemed like a long time went by before Pop said, "Who told you that?" and I didn't recognize his voice.

Mason looked like a person who had seen Death. His face was gray, even his lips.

"I asked you: Who told you that?" Pop said again. His voice was deadly. The voice of an ex-con.

Mason opened his mouth and shut it, looking strangely like a fish on land. Finally he managed, "Nobody had to tell me. I know when you went into prison and when you got out and I know his birthday. Nobody had to tell me."

Oh, Mason had gotten some weird idea.

He was Pop's kid and I wasn't. Man, he was ready for the funny farm. I looked confidently at Pop, waiting for the explanation, the real dates, waiting for him to laugh and say, "You're wrong there, Mace."

He didn't say anything. He didn't move. He didn't look at me.

The room seemed to be getting black. Time stopped, then started back up again. I tried to wake up. What was happening?

I turned to Mason.

"Tex," he said, fighting hard to sound normal, a normal voice coming from an ash-gray face. "I didn't mean . . . Texas, listen to me, kid—"

I didn't listen to him. I turned and ran instead. I didn't ricochet blindly down the hall, bouncing off the lockers like a stray bullet. I ran steadily, timing my breathing, not wild or crazy or particularly fast.

Almost like I knew where I was going. Almost like I had somewhere to go.

TEN

L

"Hey, Tex! Hey, kid, hey!"

I finally looked over at the car pulling up alongside me. I opened the door and jumped in, even before I realized it was Lem. I just wanted out of here, fast.

"Goin' somewhere?" he asked.

"Wherever you are," I said. My mind would go blank, and time would stop, then start up again with a sickening throb.

"You don't look too good, cowboy," Lem said.

"Am I awake?" I asked. It seemed like I was, but I really couldn't tell.

"Yeah. Listen, Tex, I'm headed back for the city. I got another stop to make."

"So make it," I said automatically. I was flashing hot and dizzy one second, sick and cold the next. The grass smell in the car was making me sick. I pushed the button that rolled the window down.

"I mean, I can't give you a ride back here."

"That's okay. Speed it up, willya?"

He was driving like a poke. A lot of stoned people do that. I'd rather ride with a drunk any day. If Lem didn't speed it up, I was going to reach over and bash his face in . . . I couldn't think straight. When I tried to think my mind turned into a crazy red mess.

"You know, I don't really hate ol' Mason."

Lem said it, not me. After a few minutes I realized he was continuing the conversation we'd been having that morning.

That morning was years and years ago. I could barely remember it.

"You hate a lot of people?" I asked him. The word was taking on a whole new definition for me. Like the word "water" would change for somebody drowning.

"Oh, yeah. Sure. My old man, for one. And Connie's parents for a while, they ain't so bad now. I did hate Mace some back when he made the team his sophomore year and got to thinkin' he was such a hot-shot jock and him and Bob Collins got so thick. I thought he figured I wasn't good enough to be his best buddy anymore. But now, shoot, ol' Mace would have left me behind whether Bob was there or not . . . people change, I reckon. You know hate is a real funny thing. Like George Regis, I've hated him since grade school. And the only reason I can think of is because he had a lunch pail like I wanted. Weird, huh?"

All these years and I had never hated anybody. I was dumb! It suddenly dawned on me that I must be the dumbest person in the world. I was going to make up for it, though. I was going to smarten up real quick.

Lem didn't notice the conversation being a little strange. Stoned people are always having strange conversations.

"Tex, if you're that cold, put the window back up."

"I ain't cold." I tried to stop shaking. "I ain't cold."

Mason had said, "He is my brother even if he isn't your son."

And Pop said, "Who told you that?" and that meant—oh, God, what did it mean? I couldn't think about it. I kept wiping my face off with my shirtsleeve, but the sweat kept pouring down.

"Look," Lem began slowly, "would you mind comin' in with me at this next place? I got some deliveries switched around and have to do some explainin' and it'd help if I wasn't by myself, you know?"

"All right."

"I wish you were a little meaner lookin'. You're big enough to scare somebody, though, if you just didn't look . . . who was that Connie's always saying you look like? Goofy, or—"

"Bambi," I answered absently. My fingers dug into my knees so hard the nails bit through the jeans and into the skin. I concentrated on that.

Lem didn't have anything else to say until after we'd driven through the city awhile. "Look, Tex, you just go along with anything I say, okay? If it looks like there's going to be some hassle, we'll leave. I don't want any trouble."

"You must have burned these guys bad," I remarked, trying to appear interested. I had to be normal. If I wasn't, Lem might guess . . . Lem might already know . . . who else knew . . . ? Everybody? Jamie?

For a second I almost blacked out again. I wanted to wake up, real bad.

Lem was shaking his head. "It was just an honest mistake. I can fix it."

"Don't worry about me," I said, looking out the window without seeing anything. "They try to hassle me and they'll be sorry. I'd like to make somebody sorry."

"Hey," Lam said, "I like your attitude."

By the time we pulled in the parking lot of a big apartment complex, I was really lost. I can't find my way around the city too good, anyway, even when my mind is working.

I wondered where I could go from here. Maybe Lem could use a partner. He could use some help. He wasn't exactly the smartest person in the world. I could make some money. And be doped up all the time. That sounded great. That sounded wonderful.

We went up the stairs to the second floor. There must have been a thousand apartments there. It reminded me of the time me and Mason went to see Lem. Mason never lied to me.

Sometimes it drove me nuts, but one thing I could always count on was that Mason never lied to me. He never told me the truth, either.

Lem stopped in front of an apartment that had all the curtains pulled shut. He knocked twice, then twice more. The dramatics of it vaguely irritated me. There were more important things going on . . .

We heard footsteps stop at the one-way peephole, then the door was unlocked. "We'd been wondering where you were lately. Who's your friend?"

"A connection out in Garyville. He's cool."

We went in and he chained the door shut behind us. The apartment was dark and an old Rolling Stones album was playing on the stereo. I remembered what Lem had said and tried to look mean. I couldn't concentrate on it, though. My brain was out of focus.

It seemed like right away Lem and this guy were into a heavy discussion about samples and deliveries and junk. Other than noticing that the guy seemed hacked off about something, I couldn't pay attention. It just seemed like a bunch of crap. This was not important.

I wasn't Pop's kid. That was what they were saying. My mother—why had I always thought of her as somebody really nice, somebody who would have loved me? I thought I remembered . . . but maybe I got what I remembered and what I imagined mixed up together. I had always thought she and Pop had been happy together. He never wanted to marry anybody else. And me—if I wasn't his kid, then who's? Did he even know?

It was going to be never. He was never going to care about me. It wasn't going to be next time he got back from a trip, or when Mason left or when I went on the circuit with him. There was nothing I could do to make him care.

I realized I was making a strange little sound, like a dying animal. I couldn't stop it; but Lem and his friend didn't hear.

I couldn't see how he could do that to me. Play like he cared, but not really. I couldn't help being born, I was more like him than Mason was, how could he do that? It wasn't my fault, I didn't deserve . . .

The guy Lem had been talking to started throwing a fit. He was jumping up and down, screaming, acting like he'd blown his mind. Even this took a couple of minutes to get my attention.

"I mean it man! You can't come in here giving out samples and come back with a bunch of crap! What do you think we are man, stupid? Where do you get off with this, buddy!"

Lem was stammering around, not scared, just too stoned to think clear. It annoyed the hell out of me. Mason was right. Lem was stupid. Mason was right about a lot of things. But then, he was the real kid, the one who counted . . .

"Now wait a minute, Kelly, just wait a minute," Lem mumbled.

I wasn't in the mood for this. I gave up my idea of joining up with Lem. Not if you had to put up with weirdos.

A Chinesey-looking guy came out of the bedroom. "Keep it down, Kelly, keep it down." He was almost too spaced out to talk.

"I'm leaving," I said to Lem. "These people are nuts."

I turned and started to unchain the lock.

"Where do you think you're going?" Kelly quit jumping up and down long enough to ask me.

"I'm leavin'. I don't have to take this stuff." I reached for the doorknob, but Kelly set up such a racket I turned back to watch.

"He's going to narc! He's going to narc!" He raced around the room, yanking open drawers and cabinets, feeling around under the sofa, frantic.

I was amazed. I'd never seen anybody act like that in my life. He was screaming "Narc!" till it was a wonder one didn't show up.

"No, it's cool," his friend kept saying. "It's cool."

"Don't seem too cool to me," I said. Lem nodded. "You're right. We better get outta here."

Kelly found what he was looking for. It was a .22 pistol and he had it leveled at me.

"You ain't going nowhere, man!" he screamed.

I stood there for a second, not believing it. This was the last straw. I'd put up with a lot today and this punk pulling a gun on me was the last straw. Who the hell did he think he was? A white-hot rage flashed over me and I slammed across the room at him like a bull out of a pen.

He fired once, but he didn't get a chance to fire again because I yanked that gun out of his hand and backhanded him with it. Blood gushed out of his nose as he tripped backward across the coffee table. He struggled to get back up, half caught between the sofa and the table, until he looked at me. Then he froze.

I had both hands on the gun, aiming dead on. A .22's kick ain't that bad, it doesn't need a two-handed aim, but I was shivering all over and I didn't want to miss.

His face was gray, looking at the end of the gun. I was going to kill him. I wanted that turkey dead. The trigger was warm under my finger. This was going to feel good.

"Oh, gawd, Tex, don't," Lem groaned. Out of the corner of my eye I saw him shut his eyes and pray. My heart was pounding.

The Chinesey guy was chattering, "Look man, everything is cool. Just leave. Everything is cool."

I didn't pay any attention. If he started anything I'd kill him, too. I'd like to kill him. I'd like to kill them all . . .

Kelly was too frozen to wipe the blood off his face. It seemed like I had just seen a face like that, waiting for something terrible, the color of ashes . . .

"Texas," Lem begged me, "don't do it, kid, come on, let's leave."

Not before I settled this. My finger quivered.

Then I wondered if anybody'd show up at his funeral. If he had a girlfriend, a mother, a brother. And as soon as I thought that I knew I wasn't going to kill him. So there wasn't any sense in sticking around.

"We're leaving," I said. "You try and stop us and I'll blow you away, man."

"Okay, okay, it's cool. Everything's cool." Kelly's friend went to help him up.

Lem and me backed out the door and ran down the steps. When we got to the car I realized I still had the damn gun and I threw it down the gutter.

"That is a real class set of people you hang around with, Lem. Real nice guys." My voice was shaking. Lem yanked the car into starting and squealed off.

"Man, I didn't know anything like that was going to happen! Honest, Tex, he was on something. Holy cow! Really, kid, I been doing this stuff for over a year now and I never saw nobody pull a gun before! God Almighty! What if he hadn't missed!"

"He didn't."

"What?"

"I said he didn't miss. He shot me and it hurts like hell." For a while it'd been numb. Now it wasn't numb. I never thought about bullets being hot before.

Lem looked over at me and jumped when he saw the red stain seeping through my fingers. Blood was all over. It showed up bright on the car's white interior. It hadn't been real noticeable on my navy blue sweatshirt.

"Oh, shit." Lem slammed on the brakes and stared at me. The car behind us honked loudly and we started up again with a sickening jerk. "Oh, shit. This is great, this is just great."

"What's with you?" I asked angrily. What did Lem have to be mad about? He goes for a year without any trouble and my first time helping him in the wonderful world of drug dealing, some looney shoots me.

Besides I was starting to feel funny and it scared me.

"Now I got to take you to the hospital and they'll want to ask questions and call the cops and that's all I need right now, a bunch of cops—this is just great."

We had pulled up at an intersection, but I think I would have jumped out of the car then if it'd been doing ninety down the freeway.

"Go ahead, Lem! Dump Me! Everybody else has and I don't know why you should be any different!" I was shouting at the top of my lungs. "Now get the hell out of here!" I slammed the door.

I looked around. Where did I go from here? There was a shopping center across a parking lot. I spotted a phone on a wall. I wanted to call somebody. I felt really funny. There had to be somebody to call.

I got to the phone, staggering just a little bit. Nobody paid much attention, except for a lady tightening her grip on her purse. Somebody going into the grocery store stopped, stared, and went on. I think they were used to staggering kids around there.

I had trouble getting my quarter out of my pocket. There was blood over it, all over my hand. I looked down and blood was splattering on the cement in a slow drip. I was scared. I wanted Mason. I couldn't think clear, too many things had happened too fast. Mason would know what to do. I got the quarter in the slot and realized I couldn't remember my phone number. I never called it that much. I decided to call Johnny instead. He'd know it.

"Hello?"

It was Jamie. Her voice sounded so good to me. I wished I could kiss it.

"Hey, Jamie, this is Tex."

She paused. "Yeah?"

"Yeah, listen, do you know my phone number? I can't remember it."

"You sound funny. Are you drunk?"

"No. I been shot. Listen, I want to call Mason. I got to talk to Mason. So tell me the number. It starts off three-six-six . . ."

In a very small voice Jamie said, "Did you say you've been shot?"

"Yeah, some doper friend of Lem's."

I heard her swallow. "Are you going to die?"

I thought for a minute. "Well, I don't know. I didn't think so at first, twenty-two bullets are pretty little, but I don't know, I'm feeling real strange—"

I heard her screaming. She was hollering at

somebody. Then Cole's voice said, "Is this some kind of joke?"

I stared at the phone in my hand. I started to say, "I ain't laughing," but instead I said, "No, sir." Then, "Listen, let me talk to Jamie again, okay? I . . ." I paused, trying to figure out what I wanted to say.

"Tex, where are you?"

"In town," I said, a little puzzled by why he would want to know.

"Where in town?" He sounded like he was at the end of his patience. Well, so was I! I might be dropping dead any second for all I knew and I didn't want to waste my time talking to Cole Collins.

"I'm in a shopping center," I looked around. "Down where the interstate crosses the Ribbon, you know, across from that big motel that looks like a castle." My mouth was so dry it was hard to talk. I was really thirsty. "Can I—"

"Tex, hang up the phone."

I hung it up without thinking. Then I stood and stared at it. I tried fishing around in my pockets for another quarter. I couldn't seem to get my hand in my pocket. Damn him! I felt like crying. My last quarter and I wasn't going to get to talk to Jamie or Mason or anybody . . .

Somebody tapped me on the shoulder. It was Lem.

"Hey," I said dizzily, "you got a quarter?"

I almost pitched forward. Lem caught me and propped me up against the wall next to the phone partition. "Come on kid, hang in there, we'll go to the hospital, okay? In one minute, okay?"

I blinked at the fuzzy parking lot and wondered what God was going to look like.

"Lem, would you call Mason for me? I want to tell him something." My voice wasn't much more than a whisper.

"Sure. Stay right there."

"I want a Coke," I said, yawning. I heard the coins dropping into the phone. Good. I really wanted to talk to Mason.

A couple of older ladies walked by and stared at me. I stared back. My knees started to buckle and I slid down the wall and keeled over on the cement. They went squawking off into the grocery store.

"Connie, listen, no, listen, get rid of everything. I said everything. I had some trouble at Kelly's—I know what's been paid for, get rid of it! Honey, all of it. Then go over to your sister's. Don't get hysterical, it's probably nothin' bad, but get rid of everything? Okay? I'll call you later."

Oh, God. He hadn't called Mason. Gets me shot and goes off and leaves me and now he's calling his wife while I'm probably dying.

"How you doin', Tex?" Lem stuck his head around. "Oh, shit."

He was kneeling beside me, prying my hands off my side.

A bunch of people came stampeding out of the grocery store. I hated for them to be gawking at me like that.

"Did you call an ambulance?" somebody was saying. Lem wadded up the end of my sweatshirt and pressed it over the bullet hole.

"No. Somebody go ahead and call. Texas, will you quit bleeding?"

"That hurts," I said weakly. "Hey, Lem, I want a Coke."

"Okay, just a minute. We'll get you a Coke in a minute."

I could already hear an ambulance siren. That was quick. All I could see around me was a sea of feet and legs.

"Where's Mason?" Whenever I was scared, Mason was there.

"He'll be here pretty soon. I'm going to call him, I swear."

Somebody else was shouting, "Get back, everybody back up," and someone else was putting a coat on me. I was still cold. I felt like I was laying in a cold damp cave.

"My feet are cold."

"Yeah, well you got holes in your boots."

I stared at Lem, puzzled. He wasn't making any sense.

"You better leave," I said, making an effort to think straight. "You're gonna get in trouble."

"Naw, it's cool." He dropped his voice. "Just tell them it was about a money loan, okay?"

"What was?"

"The fight."

"Okay."

My foot was going to sleep, I was laying on it funny.

"Did you call Mason yet?"

"I will, I promise."

He was getting blurrier and blurrier and I couldn't see him too good anymore.

"Tell Mason . . ." I couldn't remember what I wanted to tell Mason.

"I'll tell him. Just hang on, willya, Tex? Don't die on me, kid. Hang on."

He sounded like he was crying, but he was so far away I couldn't tell. I wished I could see Jamie. I'd give anything to see Jamie.

What was going on? This morning had been pretty normal. I wanted to ask Lem what happened, but I don't know if he heard me. I don't know if I said the words.

ELEVEN

"Mace," I said weakly, "would you get that rock off of me?"

I tried pushing it off myself, but I couldn't seem to move my arm.

"It's not a rock, it's a bandage. And don't move around, your tubes'll come loose."

I tried to focus my eyes. It seemed like I was surrounded by dangling bottles and ropes of tubes. I could barely see Mason. He looked strange.

"Are you mad at me?" I asked. I couldn't remember what had happened.

"No. Not at you. You didn't shoot yourself, did you?"

Oh, yeah. Somebody shot me . . . slowly Mason's face came into focus. He looked strange, there was something wrong . . . oh, yeah, it was a Band-Aid.

"Did somebody shoot you, too?" I whispered. It seemed like he laughed or something before he said, "No. I beat the crap outta Lem Peters—but he didn't exactly stand still and let me. Go back to sleep, Texas. They won't let me stay with you if you keep talking. You're supposed to rest."

Why would he beat the crap out of Lem? "Did he get your samples switched?" I mumbled. How did anybody expect me to sleep with this big rock on my side? It hurt like hell.

"When can I see Jamie?" I asked.

"Later."

I wanted to ask when later was, but the bed seemed to tilt and slide me right off into darkness . . .

I was miserable when I woke up next. I was tired and I hurt and I remembered what had happened. I wished I had gone ahead and died. The nurse told me I couldn't have any company except immediate family—I wasn't real sure at that point exactly who that was, but it probably wasn't Jamie.

Then they let a cop in to question me.

I did what Lem said to—told the story exactly like it happened except that Lem and Kelly were arguing over a money loan instead of a dope deal. The cop didn't seem real excited about it. I asked him if this was going to be in the paper and he said probably not, since I didn't die. Apparently kids are getting shot all the time.

That was okay with me. I was a little sick of being in the news.

"I'm going to tell you about your mother."

Pop didn't ask me if I wanted to hear it. I didn't. I thought we were doing all right, ignoring the subject. Pretending like nothing had happened. Neither Pop or Mason had mentioned why I was here or why I'd run out of the office two days ago. One of them had been here every time I could have a visitor; but the conversation never got past how I was feeling—I always said "okay" and didn't mention being tired and sore and depressed and confused. If I didn't mention it, maybe it'd go away.

"I've already told Mason how it was, and I figure the truth'll be better than what you might be thinking."

I don't know, I've managed to live without the truth all this time, I could get along without it now, I thought. But I didn't say anything.

"It was while I was in prison," he began finally.

Huh. I figured that much out myself. I glanced down at Pop. He was studying the calluses on his fingers, like maybe they'd come up with the solution for all this. He was getting bald on top. All that fun, that way he had of being happy with anything, had been knocked right out of him, like air after a stomach punch. I almost felt sorry for him.

"You got to understand that Clare was completely against that bootleggin' business I was mixed up in. She never was the kind of person who'd tell somebody what to do or not to do; she just said 'Don't expect me to be sitting twiddling my thumbs while you're sent up. If you're dumb enough to do that stuff it'll be you that pays for it, not me.'

"Well, shoot, I knew she was only part-kidding, but I wasn't figuring on getting caught. Mason had come along and we could use the money, and I was making good money—" Pop stopped, took a deep breath, and said, "—the money, hell, I've always been able to do without money. The truth is it was easy and fun and had a nice outlaw kick to it. My poor daddy, the preacher, he never understood why I got such a kick out of breaking the rules. I always did, though. Till I found out what happened to you if you got caught."

I stared at the bunch of flowers on the windowsill. My homeroom class had sent them. I had felt funny, you know, getting flowers, but I did like looking at them. If I looked at them hard enough, maybe I wouldn't hear anything.

"Anyway, I did get caught and took my rap plus a couple others my so-called friends decided to dump on me. And there was Clare, nineteen years old with a baby and me off to the state pen. When she got scared it came out mad; she always tried to hide it—being mad. I knew it, but I don't think she ever did."

"Sounds like Mason," I said. I felt real

remote. None of this had anything to do with me. None of the people in this story had anything to do with me.

"Yeah, Mason is a lot like she was, proud as Lucifer, a bulldog for grudges—she never spoke to her parents again after they tried to break us up. Bullheaded . . . you're the one that looks like her, though," Pop said simply, " 'cept her eyes were gray. She was real good with animals, especially horses. She always said they talked to her and she'd get mad when I laughed."

My wound gave a sudden throb and I had to bite back a gasp.

Then I said, "So she screwed around on you while you were in prison."

I felt Pop look at me. "I don't want to hear talk like that, Texas. I've already belted Mason once about that and I'd hate to hit a hurt kid, but you watch your mouth when you talk about your momma."

His tone was as mild as ever, but I was startled into staring at him and saw that he meant it. He'd belted Mace? He'd never hit Mason. He'd only hit me once—suddenly I remembered something else. Walking out of the police station with him and him saying savagely, "If I ever thought you was going to end up like—" that was when he hit me. And all this time I thought he meant—"If you was going to end up like me."

"I know who your daddy was." It was almost like he'd read my mind. "He was a rodeo rider;

I haven't seen him on the circuit so somebody's husband or daddy probably shot him a long time ago. Yeller-eyed tomcat, he was hanging around her, even before." The cold hate in his voice turned blood to ice. How could he hate somebody like that, all these years, and not hate me, too? "There she was, nineteen and alone and working as a waitress, while I sat up there in the pen with the worst trash that ever walked the earth due to my own stupidity. Oh, I wanted her back, even when she told me she was pregnant. She hadn't really cared anything about him, I knew that. It was just a get-even thing with me that she was sorry for, later."

Pop looked at me quickly. I didn't realize I had made any noise. "She never regretted having you, Tex, don't think that. She loved you same as she did Mason."

You could tell that half-puzzled him, even now. He continued: "So we moved up here where nobody knew us, started all over. We should have been happy. It never was the same, though. She always sort of expected me to let her down, and I never did really trust her again. That night she walked out in the snow . . . we were supposed to go to a Christmas dance, and the last minute I wouldn't go. Said I didn't feel like it, but I just didn't want other guys looking at her, dancing with her, I couldn't stand it, even though I knew I could trust her. It was like they'd know, somehow . . . she said she was going if she had to walk. She knew what I was thinking. She did walk, bullheaded . . . I followed later and we danced awhile and went over to some friends and then went home, I thought it was just a regular winter cold, the next day, and so did she . . ." He stopped. He was through.

"All this time, the way you always paid more attention to Mason than me, was this the reason?" I sounded real casual, barely curious, not like I was holding my breath. Which I was.

I wanted him to say, "No, he's just always been more trouble than you," or "No, it just seems like that with the oldest kid," or—

He said, "I reckon."

In a few minutes he said, "You look wore out, Tex. You better get some sleep."

I didn't say anything, and he left.

I did go to sleep, but first I pulled the pillow over my face and cried for a long time.

"Does it hurt much?" Johnny asked me. He was the first person to see me after they took off the family-only rule.

"Not much anymore." I was glad to see him, but I wondered where Jamie was. "When I first woke up it did. Now I'm just sore. I'm glad they took some of those needles out of me."

"The doctors say you're going to be okay. That's not what they were telling us when we got to the emergency room the other night."

"Yeah, I guess I'm lucky it didn't hit anything real important. I didn't know it was a soft-nose .22. Those bullets open up on impact and leave a pretty good-size hole. You guys were at the emergency room?" Was Jamie there, too? I wondered, but didn't ask.

"Yeah, we left for the hospital as soon as Cole called the ambulance and told them where you were."

"Oh, that was why he wanted me off the phone."

"Yeah, it's a good thing he called because you could have bled to death before Lem could get his act together and do something."

That's right, Lem had been there. "I don't remember much about being in the emergency room," I said. I could remember getting wheeled out of the ambulance and there were already tubes in my arms and bottles trailing after me like weird balloons. Somebody stuck a huge needle in my stomach, and I remembered thinking it was a strange time to be giving me a rabies shot, but a doctor told me later they were checking for blood in my stomach. There wasn't any, and apparently that was good. Everything else was fuzzy, just bright lights and people running around sticking me with needles and taking my blood pressure. Going into an elevator on a table . . .

Johnny had picked up one of my *Western Horseman* magazines and thumbed through it—suddenly he threw it across the room.

"You were asking for Jamie!" he said. "Dammit Tex, I'm the one that's been your best friend for five years! After all we've been through together, and Jamie's just a girl! We all thought you were going to die and you ask for Jamie!"

He grabbed a bunch of Kleenex off the nightstand and blew his nose.

I looked at him, miserable. And I had thought nothing could make me feel any worse.

"Johnny—"

He shook his head. "Yeah, I know, I'll understand it someday. That's what Bob says. I know this much though, some *girl* isn't going to make me forget my best friend."

I didn't know what to tell him. I reckon only people who have both been snake-bit can tell each other how it feels.

"Well, she's here," Johnny said grudgingly. I wondered guiltily if he knew how bad I wanted to ask. "I thought you'd probably want to see her alone. So I'll be going."

"Hey, man," I said, "don't go—"

He got up and finished wiping his nose. "No, I got to. We only get ten minutes apiece and Jamie'll kill me if I get one minute of her time."

"You'll be back tomorrow?"

"Yeah. Sure. Only, be careful, Tex."

"Yeah?"

"Listen, I promised God I'd never bug Cole again if you didn't die, and I don't know if I can make it."

Jamie came in. She set a box of candy on the nightstand, knocking off the Kleenex.

"Johnny helped pay for it, but I paid the most so I got to give it to you."

"Thanks," I said. I was getting turned on just looking at her.

I wondered if she could tell, but no, the blanket kept me covered up pretty good.

"I guess you'll be okay now."

"I guess so. You glad?"

"Yeah," she said, kind of defiant, "I am."

"You got pretty hysterical when I was talking to you on the phone," I reminded her.

She scowled for a second. "Well, at least I was calm when we got to the hospital. That's more than I can say for Mason or Johnny."

"Oh, yeah?" I said. Mason hadn't struck me as real shook up. "Ol' Mason get worried?"

Jamie stared at me. Her eyes were the dark blue of a flower I'd seen somewhere, a little dark blue flower with a face like a pug dog.

"Didn't anybody tell you? First Mason jumped Lem Peters right there in the waiting room and the police and orderlies had to break them up. Then, all that time you were being operated on, while you were listed as critical, he just sat and wouldn't speak to anybody. When the doctors came out and said you were going to make it, he started crying and he cried for so long the doctor gave him a shot."

"Mason went to pieces in front of everybody?" I was shocked. Mason had seemed okay when he came to visit, a little tired and quiet, but okay.

"It really was terrible. And Johnny sat there crying and Lem Peters was blubbering and I think everybody was bawling but me."

"You didn't, huh?"

"Well, at least I waited till I got home where nobody could see me."

How about that.

"Jamie, you want to see the bandage where they took the bullet out?"

When she leaned over, I slipped my arm around her and pulled her down for a kiss. She kissed me back. I felt pretty good for a sick person. After a little bit she wiggled loose.

"Ouch," I said.

"Sorry. But you were about to get me curious again."

Curious. Oh, great. I sighed. "Jamie, you think we'll ever get this thing worked out so we'll both be happy?"

"I doubt it. I don't think things ever get worked out. I do love you though. But I don't think love solves anything."

"It helps," I said. "It helps a whole lot."

The door opened. I was expecting to be a little embarrassed if it was Mason. It was Cole.

Jamie and I looked at each other, turning red at the same time. What if he'd come in a minute earlier!

"It's time to go, Jamie."

"Okay. Good-bye, Tex. See you at school."

"Bye," I said. I watched her hurry through the door. She was going to be the only girl for me. I could tell.

"I'm glad to hear you're going to be able to go home soon," Cole said.

He seemed even bigger, towering over the hospital bed. Boy, if he knew what I'd been thinking about Jamie! "Yeah," I said awkwardly, "I reckon you probably saved my life. Calling the ambulance."

I thought he might say, "Forget it," but he didn't. He said, "I hope I never regret it."

And he glanced at the door where Jamie had just been.

"No, sir," I said. When he left I understood how Johnny had felt, making a bargain with God, one you don't know you can keep.

TWELVE

"Any mail?" Mason asked me. He got home early from work on Tuesdays. He was working in a restaurant in the city.

"Just a postcard from Johnny and Jamie. They liked Disneyland and were going to see Blackie in San Francisco. I reckon him and Cole are going to bury the hatchet."

That sounded like a good way to spend spring break to me. A lot better than dragging around the house because you weren't supposed to exert yourself. At least I didn't have to stay in bed anymore.

I didn't mention the letter I got from Lem.

Dear Tex,
We decided to leave town and I wanted to tell

you I'm sorry about everything. I know Mace said he'd kill me if I ever tried to get in touch with you again. But I am sorry. Don't tell nobody where we are.

> Lem

It was postmarked Arizona.

I looked down the barrel of my rifle to make sure the sights were clean. Then I started polishing the stock.

"You going hunting tomorrow?" Mason skimmed over the postcard.

"I'm goin' for a walk and I'll take the gun with me. I ain't going to shoot anything, though. I ain't going to shoot anything ever again."

I knew how it felt.

Mason poured himself a glass of buttermilk. That was about all he could drink lately, besides water. His stomach had been acting up.

"Did you see Mr. Kencaide today?" He paused, but didn't give me the usual lecture about cleaning guns at the kitchen table. He hadn't given me the usual lecture about anything, ever since I came home from the hospital, a month ago.

"Yeah, Pop drove me over this morning. Mr. Kencaide said he could wait another week for me to start work."

"That's good."

Actually, for a minute I thought I wasn't going to get the job. Mr. Kencaide had said, "Well, I really need someone right now—"

My heart sank. The doctors had been real

positive about me not lifting or shoveling for another week. There's lots of lifting and shoveling on a horse farm.

Then he said, "Didn't I see your little brother at the Fair a couple of years ago?"

"No, sir, I don't have a little brother. But you came up and talked to me after a class once."

"Hmm. You've grown. I did like the way you handled your horse. I guess I can wait another week."

So it looked like I had a job.

Mason was shifting from one foot to another, like he wanted to say something but didn't know how to start. I went on with my gun-cleaning. I wasn't expecting him to say anything that meant anything. Ever since I came home from the hospital, we'd all been pretending nothing had happened, that the subject was dropped. Actually I don't think Pop was pretending, he was forgetting. He doesn't have a very long attention span.

But me and Mason just went on being polite to each other, which cut down on the conversation considerably. I find it real hard to live politely.

I was tired of pretending, but I didn't know how to start talking to him. Mason always has been real hard to talk to about personal stuff, and to tell somebody you still loved them was pretty personal. I didn't know if he even wanted to hear it.

"Listen," Mason said suddenly, staring hard

out the kitchen window, "I've decided not to go on to college."

I stopped, my polish rag in mid-air. I didn't say anything.

"I mean, I like working at the steak house, I've already got one raise. I can take a course in restaurant management at the junior college in the city."

For one split second I fought to hang onto that polite, impersonal ghost I'd made myself into, then I jumped to my feet and yelled, "Are you crazy?"

That startled Mason so much he jumped. But he always had had more self-control than I had, so in a minute he continued, "I've thought it all over . . ."

I've been around guns too long to be stupid enough to throw one across the room, but I felt like it. "I'll tell you what you've thought over," I said, laying the gun down carefully, because I wanted so bad to slam it into the floor. "You've thought that once you get gone Pop will leave again and I'll be on my own and I can't handle it. Well, I can. But even if I couldn't, Mason, dammit, you've tried to be my father long enough. You don't go to college because of me, and in two years you'd hate my guts."

Mason tried to go on with his calm, rational pretense, then gave up, the desperation showing plainly on his face. "I don't know what to do. I can't go. I can't stay. Sometimes I feel like I really am going to go nuts."

I knew how that felt. I knew exactly how it

felt. Late at night, lying awake, rehashing everything until my mind was whirling around like a squirrel in a cage, I thought I was going to die or go crazy. But then, in the morning, I'd still be alive, and sometimes the pain seemed a fraction less.

"I don't know what to do. Sometimes I think nothing is ever going to get worked out."

"Maybe sometimes things don't work out. I know I'll never figure out the 'why' of a lot of stuff that's happened. Mace, you never read *Smoky the Cowhorse,* did you?"

Mason was leaning back against the kitchen sink. Now he glanced over at me and grinned briefly. "No."

"Well, ol' Smoky, he had some bad things happen to him, had the heart knocked clean out of him. But he hung on and he came out of it okay. I've been bashed up pretty good, Mason, but I'm going to make it."

"You know," Mason said slowly, "I always thought—if you found out, if you knew Pop wasn't ever going to change, that he didn't care, I thought you'd—remember that hitchhiker? You said you thought something bad had happened to him. He hated everything and everybody. I thought that's what finding out would do to you. I couldn't stand that, Tex. Pop used to drive me crazy, he treated us so different, I was sure you'd start wondering why, the same as I did. I didn't see how you could help hating everybody, if you found out. Especially me."

I remembered that day, in the office. Mason's

face. The look on Mason's face before I turned and ran out. I remembered what Jamie had said, that love doesn't solve anything. Maybe. But it helps. "I don't hate you, Mace," I said.

He nodded, and all at once I realized he was crying, not making any noise, but crying.

"I know," he said, clearing his throat huskily. "At the hospital, before you went into surgery, you kept saying you had something to tell me, and when the doctors let me see you for a minute, you kept saying 'I don't hate you, Mason. I don't hate you.'"

He fell silent. Love ought to be a real simple thing. Animals don't complicate it, but with humans it gets so mixed up it's hard to know what you feel, much less how to say it. After a minute the best I could do was: "Tell you what—when you go to college, when you go out to the airport, I get to drive, okay?"

There are people who go places and people who stay . . .

"Okay," Mason said. He turned around and washed a couple of dishes. Then he said, "You need somebody to go hunting with tomorrow?"

He looked at me. And absentmindedly added, "Geez, Tex, you're getting that gook all over the table."

I started laughing. In a minute a slow grin started across Mason's face.

"Maybe we can go fishing instead," I said. Mason said okay, so we're going in the morning.